Attracting Hummingbirds and Butterflies in Tropical Florida

UNIVERSITY PRESS OF FLORIDA

Florida A&M University, Tallahassee
Florida Atlantic University, Boca Raton
Florida Gulf Coast University, Ft. Myers
Florida International University, Miami
Florida State University, Tallahassee
New College of Florida, Sarasota
University of Central Florida, Orlando
University of Florida, Gainesville
University of North Florida, Jacksonville
University of South Florida, Tampa
University of West Florida, Pensacola

Attracting Hummingbirds and

UNIVERSITY PRESS OF FLORIDA

Gainesville Tallahassee Tampa Boca Raton Pensacola Orlando Miami Jacksonville Ft. Myers Sarasota

Butterflies in Tropical Florida

A COMPANION FOR GARDENERS

Roger L. Hammer

A Florida Quincentennial Book

Title pages: ruby-throated hummingbird; *clockwise from left*, zebra longwing, queen, tiger swallowtail, gray hairstreak. *Page v*: camelia rosa; *page vi*: Atala butterfly on tropical sage.

All photographs by Roger L. Hammer

Printed in the United States of America on acid-free paper

This book may be available in an electronic edition.

20 19 18 17 16 15   6 5 4 3 2 1

Library of Congress Control Number: 2014945860
ISBN 978-0-8130-6024-8

The University Press of Florida is the scholarly publishing agency for the State University System of Florida, comprising Florida A&M University, Florida Atlantic University, Florida Gulf Coast University, Florida International University, Florida State University, New College of Florida, University of Central Florida, University of Florida, University of North Florida, University of South Florida, and University of West Florida.

University Press of Florida
15 Northwest 15th Street
Gainesville, FL 32611-2079
http://www.upf.com

To my mother, Martha,

who opened the door when I was a youngster

and let me run free in the outdoors.

Thank you, Mom.

Contents

Introduction

We've got to get ourselves back to the garden.

Joni Mitchell, "Woodstock"

The focus of this book is neither hummingbirds nor butterflies but, rather, the plants that attract them to gardens in tropical Florida. This region is blessed with pleasant year-round temperatures along with an enticing array of native and tropical plants from which to choose. Many of Florida's butterflies, like the flowering plants they visit, are tropical species that cannot survive in parts of the state that experience prolonged or hard freezes, or they are restricted in range because their native larval host plants cannot tolerate winter freezes. In addition to tropical butterflies, there are even tropical hummingbirds that occasionally cross the Straits of Florida from the nearby Bahamas, and perhaps even Cuba, to the wonderment of gardeners and bird-watchers alike. If you're lucky, one may grace your own Florida garden if only for a fleeting visit. This is not, of course, meant to belittle the charm and elegance of the ruby-throated hummingbird that embellishes gardens in tropical Florida through much of the year.

The number of plants that can be cultivated in tropical Florida is overwhelming. Not all of them are attractive to butterflies or hummingbirds, but those that do entice them—either with nectar or pollen (butterflies and hummingbirds) or as larval food (butterflies)—are worthy horticultural subjects for gardeners and nature enthusiasts alike. A landscape that has been designed and planted specifically as an invitation to hummingbirds and butterflies immediately becomes an outdoor classroom for inquisitive children as well as an endless source of soul-satisfying pleasure and rejuvenation for people of all ages. Gardens are places of peace and tranquility, but when you retreat to your garden for solitude, just glance around—you will quickly realize that you are anything but alone.

Hummingbirds and butterflies offer us a reason to plant a flowering garden; for some gardeners, they are the *only* reason. Hummingbirds are feathered jewels that almost command attention, and I never tire of watching them dart from blossom to blossom, chattering all the while. Butterflies seem less commanding, perhaps because we Florida gardeners see them nearly every day of our lives, but they still offer that closeness to nature that keeps us in touch with life itself. And who among us does not pause to admire a giant swallowtail gliding by or watch a monarch feasting on nectar from a blossom?

For as long as I can remember, I have lived among gardeners, and some of my most pleasant childhood memories were those spent with my grandfather in 1950s Orlando, Florida, where he grew rows of gerber daisies (*Gerbera jamesonii*). He grew them as a hobby, a source of income, and for contentment. I can still well remember the plants cloaked with beautiful and colorful blossoms. My grandmother had her camellias, gardenias, azaleas, and roses along with fruit trees and a chicken yard. My parents tended a shade house full of plants as well as a small water garden at their oceanfront home in Cocoa Beach, Florida, where my brother and I grew up. My brother is a passionate collector of aroids and helped found a new chapter of the American Begonia Society when he lived in Texas. Most of my friends are gardeners to some degree, and a number of them grow plants as a profession.

Over the years, my former wife, Lisa, and I managed to transform a 1¼-acre avocado grove near Homestead, Florida, into a lush oasis of native and exotic trees, shrubs, palms, flowering vines, tropical fruits, spices, and culinary herbs, complete with a natural pool where we could swim among fish and waterfalls. The considerable mischief of Hurricane Andrew aided this transformation in 1992 when the eye of the Category 5 storm passed directly over our home. Although our small 1926 home made it through the storm unscathed, the grove that once surrounded it was destroyed. After a bulldozer cleared away the debris, we had a nearly vacant acre of land on which to begin planting anew. To those of us who live in Florida, tropical storms and hurricanes are a stark reminder that even large, mature trees can be temporary residents in our gardens. But change brings renewal, and the remaking of a garden always has a way of bringing joy, pride, and accomplishment to one's life.

My wife of nearly three years, Michelle, is an impassioned collector of exotic orchids, and because of her we now have a large shade house brimming with breathtaking orchids that hail from all parts of the world.

Since I am a gardener and professional naturalist, my interests followed a natural migration toward plants that attract birds and butterflies. The excitement of seeking out new plants and then seeing birds and butterflies visit them has guided me in my life's pursuit. I regard each plant as a new companion in life, and a new plant is no different to me than a new friend. When I bring home new plants, they become my enlightenment, my inspiration, and my education.

Gardens relax us, teach us, and connect us to the soil, and being a gardener has made me a better naturalist and teacher. The comparatively small portion of the botanical world that shares our property allows me the opportunity to make observations and then teach others from personal experience. I am blessed to have had a rewarding profession as an interpretive naturalist that not only kept me immersed in nature but also allowed me to introduce others to the things I love. Also, being a gardener is deeply humbling, because you become the recipient of the friendship and generosity of fellow gardeners—and there are no souls on Earth more generous than gardeners. We all want to share our passion, so if this book helps you take on gardening as a partner in life or brings you closer to nature, then it will have achieved its purpose. Happy gardening!

Key to Symbols

= Florida native

= Attracts hummingbirds

= Attracts adult butterflies

= Butterfly larval host

= Toxic to humans and pets

Where Is Tropical Florida?

How do we define "tropical Florida," the region covered in this book? The logical line to follow is the one already created by nature, and winter temperatures are what delineate this invisible boundary. Florida lies wholly within the temperate zone, but subtropical weather prevails in the region this book considers "tropical Florida," which encompasses Hardiness Zones 10A, 10B, and 11 adopted by the United States Department of Agriculture (USDA). Officially it is called the 55° January isotherm, where temperatures within this zone average 55 degrees Fahrenheit in January.

Prior to 1980, Zone 9B encompassed much of what is now regarded by the USDA as Zone 10A, which reflects advances in technology and the current climate-warming trend. The twenty-one Florida counties that are wholly or partly within Zones 10A, 10B, and 11 are Brevard, Broward, Charlotte, Collier, DeSoto, Glades, Hardee, Hendry, Highlands, Hillsborough, Indian River, Lee, Manatee, Martin, Miami-Dade, Monroe, Okeechobee, Palm Beach, Pinellas, Sarasota, and St. Lucie. The Florida Keys portion of Monroe County is the only region that lies within Zone 11, which is shared with the nearby Bahamas and other West Indian islands.

Be aware that even in the warmest parts of Florida, there are many tropical plants that require protection from cold; some will not even tolerate temperatures in the lower 50s (Fahrenheit). Determining what to grow will to some extent depend upon your personal dedication to gardening. How badly do you want to keep your cold-sensitive plants alive through winter? Be prepared for occasional failures and losses, and then write them off as experience. There is always another botanical companion out there to fill the void.

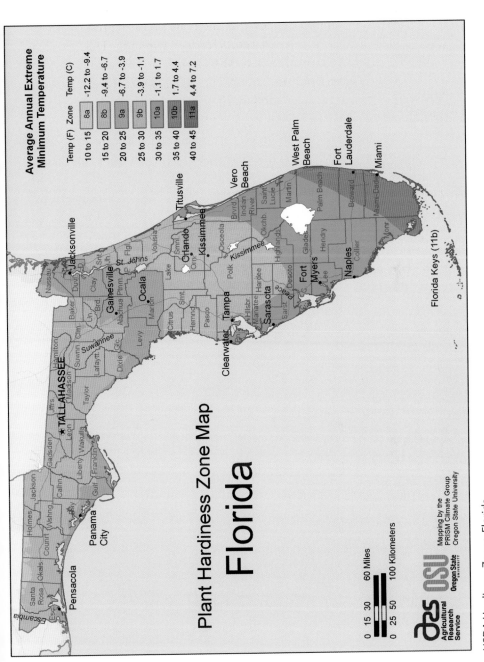

USDA Hardiness Zones: Florida.

1

The Plants

"Just living isn't enough," said the butterfly.
"One must have sunshine, freedom, and a little flower."

Hans Christian Andersen, "The Butterfly"

Pink porterweed in the author's garden is a strong magnet for hummingbirds and butterflies alike.

Whether you own acres of land, a typical urban yard, or a small courtyard, you will always be able to find the right plant for the right place to attract hummingbirds and butterflies, provided you do some simple research first. The most important thing to learn is the ultimate size of the plant you are considering for your garden. If you are holding, for instance, a sapling shortleaf fig (*Ficus citrifolia*) in a 3-gallon pot, what you really have in your hands is a tree that can reach 50' tall or more with a wide, spreading canopy. Remember this when you choose a location for trees that can grow large, and *never* consider planting a tree with the intent of severely pruning it after it matures to its natural height and spread. In this book I have attempted to give gardeners a good understanding of the ultimate size and growth habit of each species to help create an attractive and manageable garden without overcrowding. The second most important thing to remember is that if a newly installed landscape looks dense and lush from the beginning, then it is already overcrowded and will require high maintenance to keep under any semblance of control. Landscape designers and installers are notorious for crowding plants together because a beautiful finished product is what "sells" to homeowners. My best advice is for you to provide plants with adequate room and ample time to grow into their allotted space. You will then inherit a beautiful, low-maintenance and functional garden that will give you pleasure for many years.

Where you place large trees influences everything else you may want to plant in the future. If you have the luxury of being able to plan an entire landscape from scratch—that is, you have a fairly clean palette on which to design—decide where you want shade trees first. Where do you plan on spending the most time on your property, where you will want shade? Plan for some paths that lead to private, secluded areas. This can be accomplished even in small yards. If you want to attract hummingbirds and butterflies, it is imperative to have some open, sunny areas, so be sure to leave adequate room for sun-loving plants.

Avoid creating too much shade, because butterflies, hummingbirds, and the plants that attract them generally prefer protected, sunny areas. Large shade trees should be kept to the north or west of sun-loving plants.

If you have limited space, then use tall palms instead. If you do decide to use palms, avoid royal palms (*Roystonea* spp.) and other tall species that shed large, heavy fronds, because they will continually pummel everything beneath them. You might even consider trees with light, airy canopies that allow you to grow plants beneath them that prefer light shade. There are many sun-loving plants that prefer a break from midday sun.

Because of where we live, keep hurricanes in the back of your mind when designing your landscape. Large trees should be kept well away from your house so they don't come visit you in the comfort of your own home if a storm topples them. If your house is on a small lot, then plant small to medium-sized trees.

If you already have large trees on your property, then take time to assess their value. If they are undesirable species, then consider having them removed. This is especially true if you have limited space. Check local ordinances to see if a tree removal permit is required before cranking up your chainsaw or hiring an arborist. Removing mature trees may seem like a drastic measure, but there are many desirable trees that can be used as replacements, and they can even increase the value of your property. In short, do not try to landscape around an existing detriment. For further guidance, consult the Florida Exotic Pest Plant Council's list of most invasive species. If you have a listed species on your property, make removing it a top priority.

Propagation

> Gardening requires lots of water, most of it in the form of perspiration.
>
> Lou Erickson

Planting seeds is the epitome of optimism, and one of the genuine pleasures of gardening is growing your own plants. Whether you are propagating plants from seeds, cuttings, or air-layers, you will discover great pleasure and endless gratification. Here is where you might also experience failures but learn from your mistakes and move on. Some plants are much easier to propagate than others, but the tough ones can be especially rewarding when you succeed. The seeds of Bahama strongback (*Bourreria baccata*) are notoriously difficult to germinate, whereas the seeds of tropical sage (*Salvia coccinea*) sprout like radishes. Following are some tips to help with your propagation skills.

Seeds

This is the common method of propagating plants, and because a number of species in this guide are not commercially available, you will need to collect your own seeds. Be aware that collecting seeds in protected areas like parks and preserves is prohibited, so seek sources of seeds from unprotected areas, such as natural areas slated for development or areas where you have the owner's permission. It often helps to soak hard seeds in water overnight. As a general rule, seeds with hard outer coats should be *scarified* before planting: take a hacksaw blade or file and carefully cut through the seed coat. Scarifying seeds greatly increases the germination rate, especially for legumes. Other seeds may not need preparation before planting. Once your seeds are ready, fill a small pot or flat with a good potting mix or special seed-germination mix. Some plants have many small seeds; these can be scattered randomly across the soil surface and then covered very lightly with soil.

Larger seeds should be spaced evenly and planted one at a time. Don't forget to label and date them, so you will know what was planted and when. To keep the sun from bleaching away your labels, place the plastic label in the pot upside down with the plant name and date beneath the soil. And it is wise to shove the label all the way down, with just the tip showing, to keep inquisitive blue jays from pulling them all out. I speak from experience.

Some seeds germinate within several days, while others may take weeks or even months. Good soil mixes for germinating seeds can be purchased at most garden centers, but if you decide to use a seed-germination mix, which will be composed of finely ground peat and perlite, use this only in small pots, so when the young seedlings are transplanted, you can repot them in an appropriate potting mix that has better drainage. Never move seedlings or even young plants from a small pot size into a large pot. A young seedling in a 2" pot can be moved into a 4" or 6" pot but not larger. Overpotting can cause root rot due to the soil remaining wet for too long.

If you have problems with snails, or if you are growing larval host plants for butterflies, cover the young seedlings with screen to keep them from being eaten. One snail or a single butterfly larva can consume an entire tray of seedlings.

Be advised that you should *always* wear rubber gloves and a face mask when working with planting mixes that contain peat. If you have skin abrasions, scratches, or puncture wounds on your hands, a fungus called *Sporothrix schenckii* can enter your bloodstream and cause severe skin lesions called *sporotrichosis,* or rose gardener's disease. Breathing in airborne particles of peat moss can allow the fungus to infect your lungs and may result in pneumonia and tuberculosis. The bottom line is, do not let gardening be your demise.

On an environmental note, peat is the partially decomposed remains of living sphagnum moss harvested from bogs, so out of concern for the environment, we all should keep our use of peat moss products to a minimum.

Cuttings

Many gardeners believe that in order to successfully root cuttings you must have a sophisticated mist system with electric timers. This does help, but it is far from necessary. Success can be achieved without any mist system at all. Try this simple technique:

1. Fill a 4–6" pot with a good peat-based potting mix.
2. Water the peat until it is uniformly wet.
3. Take a 2–6" cutting from the plant and remove most of the leaves.
4. Dip the cut end into rooting hormone (available at garden supply stores), and shove the cutting into the soil to a depth of about 2–3".
5. If the cutting will not stand up on its own, then tie it to a small stick to keep it from moving (thin bamboo twigs work well).
6. Place the potted cutting inside a large clear plastic bag and blow the bag up like a balloon.
7. Twist the top of the bag shut and seal it tightly with a rubber band.

You have now created your own miniature greenhouse. Keep the bag under a well-lit patio or other place where it will not be exposed to direct sunlight. You may need to tie the top of the bag to something so it won't collapse onto the cuttings inside. You will quickly notice that the bag becomes coated inside with condensation, which keeps the cuttings from drying out while they produce new roots. Leave the bag alone for 3–4 weeks; then open the top slightly and let it sit a few more days so the young plant can acclimate to the drier outside air. You should now have a rooted cutting. Treat new cuttings with care until they are firmly rooted and producing new growth; then they can be either repotted or planted directly into your garden.

Air-Layers

An easy way to obtain a larger plant than you can with cuttings is to use a technique called *air-layering*. Some plants root easily in this manner, while others are impossible. Find a small branch with a diameter of about ¼–¾" that has a good shape. Take a sharp knife (there are such things as air-layering knives, but any sharp knife or razor blade will do), and after sterilizing it with rubbing alcohol, make a cut through the bark that encircles the branch 12–24" or so from the tip. Now make another cut about ½–1" below the initial cut, and finally make a vertical slit between these two cuts. Using the knife blade, peel away the bark between the two horizontal cuts. This is generally best done in spring or summer, when the bark slips away easily. Once this has been accomplished, use the knife blade to scrape the exposed area to ensure that you have entirely eliminated the connecting bark and cambium layer (the living tissue between the bark and the wood).

Next, sprinkle some rooting hormone (available at garden centers) on the cut area. Now gather some sphagnum moss, soak it in a bucket of water, and then squeeze out as much water as you can. The object is to wrap the moist sphagnum moss completely around the cut on the branch. After enclosing the cut with the sphagnum moss, wrap it tightly with aluminum foil, twisting each end firmly around the branch to completely seal the moss inside the foil. As a precaution, it is also wise to wrap the aluminum foil with burlap or some other bland-looking cloth to keep inquisitive birds from pecking through it. Now simply wait about a month, then gently peel away the foil to check for roots, and if they are present, take pruning shears and cut the branch off just below the sphagnum ball and remove the foil. You can now gently pot the rooted branch with a stake to hold it firmly upright (do not remove the sphagnum moss from the roots before potting). Once you see roots at the bottom of the pot through the drain holes, it can be moved to its permanent place in your garden, or repotted.

Air-layers and cuttings have the advantage of allowing you to acquire a genetically exact replica of the parent plant. If you find an exceptionally superior plant of a particular species, air-layers or cuttings are the way to go. Also, if the plant is dioecious, meaning male and female flowers are produced on separate plants, air-layers and cuttings allow you to select the sex of your plant. This is not an option with seeds.

Containerized Plants

A common horticultural mistake is keeping woody plants in small containers for too long—this is especially true for large trees and woody shrubs. When roots grow to the bottom of a container, they are forced to follow the curvature of the container or escape through the drain holes and root into the ground. Because most containers are round, the roots begin to grow in a circle and will eventually encircle the inside of the container several times. This process is called *root-binding,* and it is very difficult to remedy. The key is to never let it happen in the first place. Trees that are root-bound can be planted in a landscape and survive for years, but they will tend to remain stunted and can topple easily in strong winds. The roots may eventually constrict so tightly that the tree dies.

If you are purchasing potted plants from a nursery, remove the plant from the container and inspect the roots to ensure that large roots are not encircling the root-ball. Also avoid plants that have had large roots cut off after they have grown out of the drain holes into the ground. A container-ized plant should be vigorous, with roots just reaching the outer confines of the container. Before planting, gently pull some of the main roots away from the root-ball to encourage them to grow outward once the plant has been placed in the ground.

Root-binding does not affect herbaceous species and palms as much as it does trees, shrubs, and large woody vines, but it is always a good idea to slightly loosen the soil around the outside of the root-ball to encourage lateral root growth.

Mulch

Most plants benefit from a layer of mulch over their root zone. This helps maintain soil moisture and also helps to cool the soil over the roots, controls weeds, slowly leaches nutrients into the soil, and helps prevent mechanical injury to the trunk and surface roots caused by mowers and weed trimmers. When placing mulch around plants, ensure that it is not mounded against the trunk, because this can cause basal trunk rot. Spread the mulch to a depth of 2–3" evenly over the ground to a distance equal to the canopy spread if possible.

Because cypress mulch is no longer a by-product of harvesting trees for lumber, environmentally conscious gardeners should choose another type of mulch. Mulch that has been color dyed should also be avoided if it has been derived from recycled wood or pressure-treated wood. Some brands of red mulch may leach arsenic into the soil (some of this "designer" mulch is made from pressure-treated lumber, which contains 22 percent pure ar-senic derived from chromated copper arsenate used in the treatment of the wood). Because of the health hazard from arsenic, it is especially danger-ous to use where children play.

Eucalyptus mulch is excellent because this tree is a sustainable crop, and it actually repels some types of insects. Commercial melaleuca mulch is another good choice because cutting down the trees for mulch helps rid Florida of this invasive pest species, also called cajeput. Its seeds will not germinate unless you spread them in a freshwater wetland. Some gardeners even choose to purchase their own chipper to help recycle small twigs and branches that are pruned from plants on their own property.

Leaf mulch breaks down into the soil most rapidly and therefore benefits the plant sooner, but it lasts the shortest amount of time. Pine bark nuggets, on the other hand, decompose very slowly, and some gardeners choose to scatter pine bark nuggets on top of mulch because of the aesthetic appeal. Chipped pine bark and pine bark nuggets also help acidify the soil, which may benefit some plants growing in heavy limestone soils.

Fertilizer

There is a common mistaken belief that Florida native plants do not need to be fertilized—it is more accurate to say that many Florida native plants do not become nutrient deficient as often as some exotic, non-native species. This is principally due to Florida's native plants being better adapted to the soil types available to them within their natural habitats.

However, just as you might take vitamin supplements as a sound practice for good health, regular applications of fertilizer will benefit Florida native plants as much as they do any other plant. Waiting for a plant to become nutrient deficient before giving it fertilizer is no wiser than waiting to take vitamin C until you get scurvy. Regular applications of fertilizer help avoid nutrient deficiencies over the long term, so I recommend light applications of fertilizer on a regular basis. Top dressings of kitchen compost always is a good practice, and organic products like fish emulsion can be used for herbaceous species or potted plants. There are slow-release fertilizers suitable for potted plants as well.

Different soil types require different fertilizer mixes, so check with your local Agriculture Extension Service or a horticultural consultant for advice on the best mix for your region.

Frost Protection

There will probably be a few winter days and nights where protection from frost or freezing temperatures is necessary. This can be accomplished in a number of ways. Most gardeners bring cold-sensitive potted plants indoors until the threat of frost has passed. If that is not possible, then crowd the plants together and cover them with a sheet or blanket. *Do not use plastic sheeting* unless it is held above the plants and sealed around the edges to create a greenhouse effect. Allowing the plastic to touch the plants will transfer the cold to the leaves and stems. The thin blankets that moving companies use to protect furniture work well, as do painters' drop cloths, which come in various sizes and are available from hardware stores. Old sheets and blankets from your closet work perfectly well, too.

There isn't much you can do about protecting trees, but if you have cold-sensitive shrubs in your landscape that are too large to be covered with a blanket, then pile mulch against the trunk. If the top of the shrub is killed, then at least you will have saved the roots and lower trunk so it can resprout in springtime. If you know it's going to freeze, another option is to run your sprinkler over shrubs all night long, which will coat them with ice. As odd as it sounds, the ice will protect the plant from exposure to temperatures below freezing.

Always wait until new growth has begun to emerge in spring before cutting off the dead stems. The new growth will let you know exactly where you should prune.

A Word About Natives Versus Exotics

> Horticulture—a minor industry in terms of its economic size—is a gargantuan engine of biotic mixing that has helped unleash some of the world's worst plant invasions.
>
> *Chris Bright,* Life Out of Bounds

The debate over the virtues of native plants versus exotic (non-native) plants has tempered in recent years. A movement toward planting native species has become a genuine passion for many people across the nation. Florida is at the forefront, because our state has suffered extraordinary habitat loss from invasion by exotic pest plants. Likewise, our food crops and landscape plants are threatened by a host of exotic animals and even foreign diseases and pathogens.

But the habitat loss is deeply concerning. Because some exotic plants have proven to be overly aggressive in their ability to invade and adversely alter natural ecosystems, these plants need to be eradicated, and that eradication should begin in your home garden. If a plant becomes too weedy, then do yourself and the environment a favor by removing it permanently. Life is too short to be troubled by weedy plants, and a sunny spot in a garden is too valuable to allow it to be taken over by a pest. Take responsibility. Do not cultivate or disseminate plants that invade natural ecosystems.

Native plants are not inherently "better" than exotic plants, although they do have many virtues. Florida's native plants all arrived here by natural means (wind, ocean currents, migratory birds), and those that became established before the first Spanish galleons arrived at our shores are naturally adapted to Florida's soils, temperatures, and weather patterns. Many exotic plants have special horticultural requirements that are sometimes difficult or even impossible to meet in Florida. For instance, high-altitude plants often do not fare well when planted near sea level. Plants that prefer acidic soil are intolerant of the alkaline soils that typify parts of Florida, and those species that hail from arid regions quickly suffer in our rainy, humid climate. Cold-hardy temperate plants will likely wilt away in our summer heat, while some tropical species are either severely pruned or killed by occasional frosts and freezes. Vegetable gardens are generally planted in fall and winter in Zones 10 and 11, which aptly demonstrates how different gardening is in this region compared to elsewhere.

Native plants offer natural food and shelter for native wildlife. This is not to say that exotic plants will not attract wildlife, because this is clearly to the contrary. Your choices of plants that attract butterflies and hummingbirds in tropical Florida will be far greater with exotic plants than with native plants—but this is due only to the sheer numbers of cultivated exotic plants from all parts of the world. But many of the choicest butterfly-attracting native plants are not normally cultivated. This does not mean they do not merit a space in your garden; it only means that nursery growers, for whatever reason (demand, seed source, lack of knowledge), do not propagate them. This is when the fun begins, because you can collect your own seeds or cuttings of these species and grow something unusual. And you will have lasting memories of when and where you collected seeds of your prized plants.

Gardens that are created wholly of Florida native species can be as attractive as any other landscape design. I hold a special fondness for Florida's native plants, but my passion for gardening, and for the unusual and bizarre, has led me to love many exotic plants as well. Still, native plants are the best core for landscaping. Henry Nehrling, a renowned early twentieth-century Florida gardener, botanist, writer, and plant collector, said it most succinctly: "The native plants should form the foundation of every American garden." I wholeheartedly agree.

This book offers advice regarding both native and exotic plants, but if a plant has any horticultural or environmental concerns, they are duly noted. To start, listed below are some species (with botanical synonyms to avoid confusion) that are sometimes suggested in books on butterfly gardening that should be avoided, due to their aggressively weedy tendencies in garden settings and their ability to escape cultivation and adversely affect natural areas:

Phillippine violet (*Barleria cristata*)
Rose glory bower (*Clerodendrum bungei*)
Java glory bower (*Clerodendrum speciosissimum*)
Lantana (*Lantana camara*)
Twin passionflower (*Passiflora biflora*)
Mexican flame vine (*Pseudogynoxys chenopodioides; Senecio confusus*)
Mexican petunia (*Ruellia brittoniana; R. simplex; R. tweediana*)

A Cautionary Word About *Lantana camara*

The very popular *Lantana camara* is not recommended in this guide because it is weedy, it invades natural areas, it hybridizes with a Florida

Lantana camara, a Category I invasive species.

endangered, endemic species (*Lantana depressa*), the green, unripe fruits can be fatal to children if eaten, it is highly toxic to grazing livestock, and it can give your dog severe liver problems. Yes, it is exceptional at attracting butterflies and hummingbirds, and it thrives in poor soils where other plants fail, but it is listed by the Florida Exotic Pest Plant Council in Category 1 of Florida's most invasive species. It is banned in many African countries because it kills more cattle than any other plant. A wiser choice would be to grow the Florida native wild sage (*Lantana involucrata*) treated in this guide. Another option is to select from the many sterile hybrids of the Lantana Callowiana Hybrid Group in the nursery trade. These include 'Dwarf Pinkie', 'Gold Mound', 'Lemon Swirl', 'Lola', 'New Gold', 'Texas Flame', 'Weeping Lavender', and 'Weeping White'.

2

Species Accounts

A tiger swallowtail nectars on the flowers of a Florida native blazing star.

For convenience, the plants in this guide are arranged by growth habit (trees, shrubs, vines, herbaceous species, and groundcovers). Within these categories the plants are arranged alphabetically, first by genus and then by species. Botanical names are the valid names of plants as opposed to common names, which vary from one region to the next or even within the same region. Common names used in this guide are mostly those that are in common usage in Florida; they can be referenced in the index along with the botanical names. The Latinized botanical names in this guide mostly conform to The Plant List, which is a working list of all known plant species, available on the Internet, created by the Royal Botanic Gardens, Kew, and the Missouri Botanical Garden in collaboration with many other institutions and databases. The Plant List provides the accepted Latin name for nearly 300,000 plant species with links to synonyms. These names are also cross-referenced with links to other institutions and databases.

To become a knowledgeable gardener you should pay attention to plant families and Latin names, so you can better understand their relationship to other plants in your garden. Moreover, to reference a plant in various books or on the Internet, you will have much better luck finding the correct plant by using the Latin name. Even Latin names can change on occasion, especially with recent studies of plant relatedness using DNA, so synonyms are listed for species that have undergone relatively recent name changes. This will help you to research the correct plant by using an older Latin name that still appears in many publications and Web sites.

The main purpose of this book is to offer accurate information on 200 of the best plants that attract hummingbirds and butterflies in tropical Florida. Being educated about what you are growing will make you a better gardener.

Large Trees

This section includes trees that typically reach heights of 30 feet or more. Put some thought into where they are planted, so they will not create too much shade for sun-loving plants.

Silk-floss tree, *Ceiba speciosa.*

Hong Kong Orchid Tree

Bauhinia × *blakeana*
Fabaceae (Pea Family)

Flowering season: Fall, winter, spring.

Native Range: A sterile hybrid discovered in Canton, China.

Comments: This is unquestionably one of the premier exotic flowering trees in tropical Florida. Besides being glamorous, it is a sterile hybrid that cannot produce seeds and become a weedy pest like other members of this genus. The rose-purple flowers reach 4–5" wide and appear in showy clusters during the peak of winter. It is deciduous and typically drops its leaves in late spring, so plant it where leaf litter will not be a problem.

This is a favorite nectar source for hummingbirds. When in bloom, these trees are often surrounded by the chattering of these birds as they defend this prized territory from intruders. The fragrant flowers also attract butterflies as well as nectar-seeking orioles and warblers.

It can reach 25–30" tall, requires full sun for maximum flowering potential, and is best used as a centerpiece in the landscape. Two related species, *Bauhinia purpurea* and *B. variegata*, are especially invasive, so do not be enticed by their pretty blossoms.

Silk-Floss Tree

Ceiba speciosa

Bombacaceae (Bombax Family)

Synonyms: *Chorisia speciosa.*

Flowering season: Fall into spring.

Native Range: Brazil and Argentina.

Comments: The striking flowers of this tree create a flamboyant display beginning in fall and lasting into winter. The tree can reach large proportions, often 40' or more in height, with numerous sharp, stout spines covering the trunk and branches. The leaves drop by early spring, but new growth often appears as old leaves are still falling. Hummingbirds and skippers gather around flowering specimens, and in springtime free-flying wild parakeets rip open the green pods to eat the seeds, scattering white fluff like snow.

The silk-floss tree is hardy into Zone 9B but is more commonly seen in southern Florida. The eye-catching, hot pink flowers measure 4–5" wide and literally consume the canopy. When in flower it rivals any other tree in beauty.

There are several cultivars available that have superior traits or unusual flower colors and will bloom at a young age because they are grafted. The Spanish name *palo borracho* translates to "drunken stick" and may relate to its use in the Amazonian hallucinogenic drink *ayahuasca.*

Colville's Glory

Colvillea racemosa

Fabaceae (Pea Family)

Flowering Season: Fall.

Native Range: Madagascar.

Comments: When you first lay eyes on Colville's glory in flower, you will need to sit down and catch your breath, because it is one of the most beautiful trees in the world. Explaining its elegance requires descriptive adjectives like *flashy, elaborate, glorious, stupendous,* and even *baroque.* It is simply stunning. Bees, butterflies, hummingbirds, and orioles visit the flowers, which hang down in foot-long, extraordinarily beautiful racemes.

Although it can grow taller, it tends to top out at about 30' in Florida with a relatively open canopy of airy compound leaves, much like the related and overly popular royal poinciana (*Delonix regia*), also native to Madagascar.

Colville's glory must have full sun and fully deserves a high-profile, front-and-center position in your landscape. It is available from specialty nurseries in southeastern Florida or perhaps from plant sales held at botanical gardens. Internet sources also offer it for sale. It is hardy in Zones 10B and 11 but may survive in coastal regions of Zone 10A.

Shortleaf Fig

Ficus citrifolia
Moraceae (Mulberry Family)

Flowering Season: All year.

Native Range: Southern Florida, Bahamas, and West Indies to the American tropics.

Comments: This is a Florida native tree of grand stature, reaching 50′ tall with a stout trunk and spreading canopy. It is a first-rate shade tree and one of the best bird-attracting native trees in all of Florida. Blue jays, white-crowned pigeons, northern mockingbirds, and cedar waxwings are among the birds that savor its small figs. Plus, warblers, gnatcatchers, vireos, and flycatchers flit around in the canopy during migration as they seek insects.

As if its power to attract birds was not enough, the leaves are an important larval food for the ruddy daggerwing, a fast-flying tropical butterfly seen in urban gardens from lower Central Florida southward.

Give shortleaf fig the room it requires. Also, because of its aggressive roots, keep it well away from foundations, driveways, septic tanks, and drain fields. It is not at all suitable for a small property. The Florida native strangler fig (*Ficus aurea*) can be substituted, but be prepared for seedlings to show up as epiphytes on other trees, palms, and rain gutters. Avoid large non-native *Ficus* because many are invasive and tend to topple in storms.

Wild Tamarind

Lysiloma latisiliquum
Fabaceae (Pea Family)

Flowering Season: Spring into fall.

Native Range: Southern Florida, Bahamas, Cuba, Mexico, and Belize.

Comments: If you are looking for a large, fast-growing, Florida native tree that attracts more than its share of birds and butterflies, then put wild tamarind at the top of your list. Its typical mature height is 30–35', but it can grow much taller in forests. It is an exemplary bird attractor, especially for migrating warblers, gnatcatchers, vireos, and flycatchers. Butterflies visit the flowers, which look like little powder puffs, and the larvae of the mimosa yellow, large orange sulphur, Cassius blue, and amethyst hairstreak butterflies feed on the leaves. There are also species of small moths that use the leaves as larval food; their caterpillars are what attract so many insect-eating birds.

The small leaflets give the spreading canopy a light, airy appearance and allow plenty of light to filter through to understory plants. The leaves are deciduous, falling in late winter and spring, with new leaves emerging soon after the first spring rains. It gets a resounding recommendation and is readily available in nurseries that specialize in Florida native trees.

Horseflesh Mahogany
Lysiloma sabicu
Fabaceae (Pea Family)

Flowering Season: Spring into fall.
Native Range: Bahamas and West Indies.
Comments: This tree is not related to the true mahogany, but the wood is deep red, hence the reference to horseflesh. It is an elegant, 30–35' tree with vertical strips of bark along the trunk and a graceful, weeping canopy. Following winter dormancy, the entire tree flushes with bright red new growth.

Horseflesh mahogany has recently become the center of attention in South Florida due to the sudden revelation that pink-spot sulphur butterflies were among us. This bright yellow butterfly has been in Florida longer than anyone knew, but it was only discovered in 2012. The leaves of horseflesh mahogany are its larval food, and surveys conducted in South Florida in 2012 revealed that pink-spot sulphurs were present wherever this tree occurred.

The butterfly was discovered when a lepidopterist reviewed pinned specimens at the University of Florida and noticed a tiny pink spot at the base of the wings on a few specimens in a collection of similar Statira sulphurs. That discovery caused butterfly enthusiasts to fan out across South Florida in search of pink-spot sulphurs. As it turns out, they were right before our eyes all along.

Live Oak

Quercus virginiana
Fagaceae (Beech Family)

Flowering Season: Spring.

Native Range: Southeastern United States south through the Florida mainland west to Oklahoma and Mexico.

Comments: Although the noble live oak is seldom regarded as a tree that attracts hummingbirds or butterflies, it is both. Hummingbirds are very fond of oak pollen, and if you find a live oak in flower you will likely see or hear hummingbirds in the canopy. Also, the white M hairstreak, Juvenal's duskywing, and Horace's duskywing, along with some of Florida's prettiest moths, all use the leaves of live oak as larval food.

Live oak requires spacious room for its spreading canopy. Although it has a mediocre growth rate, it is a grand tree that should be honored as such in the landscape. It is not a tree to be kept in bounds by pruning, so do not plant one if it cannot grow to its full majestic potential. Squirrels and many species of birds, especially woodpeckers, are attracted to live oak, and it is a superb tree to decorate with orchids, bromeliads, ferns, and other epiphytes. It is called live oak because it retains its leaves in winter, when other oaks, having dropped their leaves, look dead.

Medium to Small Trees

This section includes trees that typically range from 12 to 25 feet tall at maturity. Some can kept smaller by pruning if necessary.

Peregrina, *Jatropha integerrima*.

Pata de Vaca

Bauhinia divaricata
Fabaceae (Pea Family)

Flowering Season: All year.
Native Range: Greater Antilles, Mexico, and Guatemala.
Comments: This pretty tree was introduced into the Florida nursery trade in the early 1950s from both Cuba and Jamaica but surprisingly never gained popularity. It is exceptionally cold hardy despite its natural range in the tropics and has survived temperatures down to 10°F in Texas. The showy flowers are produced in short, elongated racemes, open white, and usually turn pink with age. The flowers open randomly, so there are almost always white and pink flowers present.

The tree reaches about 16' tall, so it has potential for use as a colorful landscape tree for small yards or where space is limited. It requires full sun for optimum flowering, and it benefits from occasional pruning to maintain a more compact canopy.

Besides hummingbirds, a variety of butterflies visit the flowers, especially during the sunny days of summer. Sulphurs, swallowtails, and skippers are frequent visitors by day and sphinx moths sip nectar from the blossoms during the evening. It is occasionally used as a street tree in the Miami area, where it is called *pata de vaca*, Spanish for "cow foot," in reference to the leaf shape.

Bahama Strongback

Bourreria succulenta

Boraginaceae (Borage Family)

Synonyms: *Bourreria ovata*

Flowering Season: All year.

Native Range: Southern Florida, Bahamas, and Greater Antilles.

Comments: Weeping branches lend style and grace to this Florida native tropical tree. Among its many redeeming qualities, the clusters of small, fragrant flowers attract an impressive assortment of butterflies and are a favorite of hummingbirds as well. Northern mockingbirds, gray catbirds, and blue-headed vireos savor the small, ornamental, orange fruits while warblers, vireos, gnatcatchers, and flycatchers search the canopy for small insects attracted to the blossoms. Talk about multitasking.

This tree takes up little space and can be used as a freestanding specimen or to create a tall screen in full sun or light shade. In nature, it typically occurs along coastal forest margins, especially in the Florida Keys, where it reaches about 14–16' tall at maturity. The very similar rough strongback (*Bourreria radula*) has rough, sandpapery leaves and is very rare in cultivation. The smaller pineland strongback (*B. cassinifolia*) is in the Shrubs section of this guide.

The name *strongback* (often corrupted as "strongbark") relates to its use as a tea in the Bahamas "to give men a strong back," with sexual implications. It can be found in nurseries that specialize in native plants.

Pride of Barbados

Caesalpinia pulcherrima
Fabaceae (Pea Family)

Flowering Season: Spring into fall.

Native Range: West Indies and tropical America.

Comments: This is the national flower of Barbados, and a pair of blossoms adorns Queen Elizabeth II's Barbadian flag. To add to this tree's stunning beauty, giant swallowtail and Gulf fritillary butterflies love the flower nectar as much as hummingbirds do.

Before running off to the nearest nursery, do yourself a favor and seek out the thornless form of this species; otherwise, you must deal with some wickedly sharp spines and recurved thorns produced along the branches and leaf petioles. This tree is positively flamboyant when in flower, which is frequent throughout the warmer months in Florida. It matures at about 10' tall and is usually deciduous.

Beanlike pods enclose toxic seeds, so keep them away from children. The pods can be removed and discarded, or you can allow the seeds to mature for gardening friends to grow.

The thornless, user-friendly form is an outstanding tree for small yards, and the lacy foliage offers the perfect shade for smaller plants that require bright light but not full sun.

Red Powderpuff
Calliandra haematocephala
Fabaceae (Pea Family)

Flowering Season: All year.
Native Range: Bolivia.
Comments: Red powderpuff is familiar to most Florida gardeners. It was once very popular but has gone out of vogue in recent years. It matures at about 12′ tall with a spreading canopy that exceeds its height. The ornamental flower clusters with their long red stamens resemble powder puffs and are highly attractive to hummingbirds, tanagers, orioles, and some of the larger species of butterflies. The leaves even serve as a favored larval host of the Statira sulphur butterfly in southern Florida.

Red powderpuff will take up an area 12–16′ wide, so be certain to allow sufficient room for it to spread. Young plants are commonly offered at nurseries that specialize in flowering trees and are sometimes available in mainstream garden centers. Annual applications of fertilizer and chelated iron will alleviate nutrient deficiencies.

In the landscape, it looks best as a centerpiece surrounded by open lawn to show off its growth character and cherry-red flower clusters, which decorate the branches most of the year. Pink powderpuff (*Calliandra surinamensis*) is practically identical but has pink flowers.

Seven-Year-Apple

Casasia clusiifolia

Rubiaceae (Madder Family)

Synonyms: *Genipa clusiifolia*.

Flowering Season: All year.

Native Range: Coastal southern Florida, Bermuda, the Bahama Archipelago, and Cuba.

Comments: This small tree matures at about 8–12′ tall and merits greater attention by Florida gardeners. The very ornamental, glossy, dark green leaves are produced near the branch tips with perfumed, tubular, white flowers emerging from the leaf axils. Male and female flowers are on separate trees, and the females produce somewhat edible fruits that, despite the common name, take about 7 months to ripen, not 7 years. When the fruits are soft and wrinkled, bite off the stem end and squeeze the black pulp into your mouth. It tastes a bit like licorice.

Because it is an inhabitant of maritime forests, it is a good choice for landscaping where salt tolerance is a requirement. In the wild it is often found as an understory tree, but in the landscape it should be given ample sunlight for best growth character and flowering potential. Hummingbirds and butterflies seek nectar from the flowers.

Your best source will be nurseries that specialize in native plants. If you want fruits, you will need to buy a male and a female. It was recently moved back into the genus *Casasia* from *Genipa*.

Golden Shower

Cassia fistula
Fabaceae (Pea Family)

Flowering Season: Spring into fall.
Native Range: India.
Comments: This popular flowering tree puts on a sensational display of pendent, 12–18" racemes of yellow flowers, often when the tree is leafless in spring. It is a street tree in Miami and is also popular in home landscapes throughout southern Florida. One horticultural concern is leaf yellowing from iron deficiency, but this is easily remedied by adding chelated iron as a supplement to normal fertilizing. Yellow leaves with green veins are diagnostic of iron deficiency. This tree will reach about 25' tall but blooms when relatively young.

Golden shower is a larval host plant of the orange-barred sulphur, cloudless sulphur, Statira sulphur, and sleepy orange butterflies. Feeding damage from their larvae may be noticeable, but this should not preclude using it as a focal point in the landscape. Remember that it is briefly deciduous in late winter to early spring, so place it where leaf fall will not pose a problem. Other choices are the pink shower (*Cassia bakeriana*), apple blossom cassia (*C. javanica*), and rainbow cassia, a charming hybrid between the golden shower and apple blossom cassia that displays pendent clusters of yellow flowers with pink highlights. 🦋 🐛

Mayan Spinach

Cnidoscolus aconitifolius
Euphorbiaceae (Spurge Family)

Synonyms: *Cnidoscolus chayamansa.*
Flowering Season: All year.
Native Range: Tropical America.
Comments: Mayan spinach is a small, semiwoody tree that reaches about 12' tall with dark green leaves that are usually deeply dissected. Erect flower spikes reach 6–12" tall, are branched at the top, and have small, white flowers that provide nectar for virtually every species of butterfly within fluttering distance. In Mexico and Central America, the young leaves are cooked as greens and are high in minerals and carotene. This plant can be propagated with ease by simply cutting off a woody stem and sticking it in a container filled with potting soil. It roots so easily that the leaves of cut stems usually do not even wilt or fall off.

In full sun, Mayan spinach produces a dense, umbrella-shaped crown topped with flowers throughout much of the year. It can be found in nurseries that specialize in edible plants or at plant sales held by the Rare Fruit Council International or by similar groups. Events held at the Redland Fruit and Spice Park in southern Miami-Dade County are a source of many unusual, hard-to-find edible plants from the tropics. Or, if you know someone who has one in their landscape, ask them for a cutting to root. 🦋

Geiger Tree
Cordia sebestena
Boraginaceae (Borage Family)

Flowering Season: All year.

Native Range: Coastal southern Florida, West Indies, and tropical America.

Comments: The Geiger tree is a fashionable flowering tree in southeastern Florida and the Florida Keys. It is almost certainly a native of coastal habitats in Miami-Dade and Monroe counties, but this is disputed. The dispute is due to reports of it being cultivated around Key West in the 1830s and not recorded as a wild tree by early botanists. The fruits float in seawater and are consumed by migratory white-crowned pigeons, so it likely arrived in Florida by natural means long before the arrival of Spanish galleons. If so, then it is a Florida native tree.

Regardless of its unresolved native status in Florida, it is exceptionally attractive when in flower. It is, however, cold sensitive, and a hard freeze can kill mature trees. Also of concern are the hideous-looking larvae of a tortoise beetle that can render the leaves unsightly. The mature beetles are gorgeous, if that is any consolation. It reaches 10–14' tall, and hummingbirds and butterflies savor the showy clusters of orange to orange-red flowers.

The famous ornithologist John James Audubon named the tree to honor Captain John Geiger, a sailor who introduced the tree to Key West and whose home is now Key West's Audubon House.

Lignumvitae

Guaiacum sanctum

Zygophyllaceae (Caltrop Family)

Flowering Season: Mostly springtime but sporadically several times a year.

Native Range: Florida Keys and tropical America.

Comments: Lignumvitae is among the royalty of native tropical trees in Florida. A mature specimen exudes character with its tidy canopy of dark green leaves covered with small blue flowers. The flowers are followed by fruits that turn yellow before splitting open to reveal hard, red-coated seeds. Birds consume the seeds while skipper butterflies frequent the flowers. The tender young leaves serve as larval food for the lyside sulphur, a rare butterfly of the Florida Keys that occasionally strays onto the southern Florida mainland.

Lignumvitae deserves a dignified place in your landscape, either formally on either side of an entryway or as a freestanding centerpiece. Its mature height is about 8–10', but it can grow taller.

Although propagation is usually by seed, lignumvitae is notoriously slow-growing, adding credence to the old saying that the best time to plant a tree is "twenty years ago." Potted saplings and field-grown specimens are available in southern Florida. The common name means "wood of life," referring to the imagined magical properties of the resin.

Mexican Firebush

Hamelia patens var. *glabra*
Rubiaceae (Madder Family)

Synonyms: *Hamelia nodosa.*
Flowering Season: All year.
Native Range: Southern Mexico through Central America into northern South America.

Comments: This non-native variety of Florida firebush is included here more as a warning than a recommendation to native-plant enthusiasts because it is often erroneously sold by nurseries as a Florida native. It was first introduced in the mid-1980s from a botanical garden in Pretoria, South Africa, and given the inaccurate and misleading common name, "African firebush." Cultivar names include 'Dwarf,' 'Compacta,' and 'Macrantha.'

Mexican firebush (var. *glabra*) differs from Florida firebush (var. *patens*) with its yellow, orange-based flowers and smooth leaves arranged in whorls of 4 along the stem. Butterflies and hummingbirds swarm around the flowers.

Mexican firebush reaches 10' tall with a rounded canopy, but it is seen in mass plantings along roadways of southern Florida where it is kept low by hedging. It can hybridize with other members of the genus in Florida, which is a concern for wild populations of Florida firebush. The misnamed Bahama firebush (*Hamelia cuprea*) from the Greater Antilles has bell-shaped, yellow flowers with orange stripes; although it is exceptionally pretty, hummingbirds and butterflies mostly ignore it.

Florida Firebush

Hamelia patens var. *patens*
Rubiaceae (Madder Family)

Flowering Season: All year.

Native Range: Central and southern Florida through tropical America.

Comments: Florida firebush is an absolute must for any hummingbird and butterfly enthusiast in tropical Florida. If you are starting with a clean palette, install Florida firebush first and then figure out what else you want. It is one of our most colorful, productive, and well-known native shrubs, providing year-round color and an overabundance of activity from nectar-seeking hummingbirds and butterflies as well as songbirds that feast on the fleshy purple fruits. Sphinx moths visit the flowers after sunset, and the leaves are food for larvae of the Pluto sphinx, an attractive green moth with swept-back wings, and the brown tersa sphinx.

Florida firebush may reach 16' tall but will more typically be a 6–8' shrub. Colorful clusters of reddish-orange, tubular flowers are present every day, and when grown in full sun, the leaves will be tinged with red. It is widely cultivated in central and southern Florida, especially by butterfly gardeners and native-plant enthusiasts. The key to identifying Florida firebush is its red or reddish-orange flowers with slightly hairy leaves mostly arranged in whorls of 3 along the stems.

Peregrina
Jatropha integerrima
Euphorbiaceae (Spurge Family)

Flowering Season: All year.

Native Range: West Indies.

Comments: Peregrina is a small tree that matures to a height of about 10', but it produces flowers at a very early age. The flower spikes branch at the apex and produce an abundance of coral-red blossoms throughout the year. There is also an attractive pink-flowered form in cultivation, but it is much less popular. Both color forms attract an assortment of nectar-seeking butterflies, and ruby-throated hummingbirds occasionally visit, too.

Peregrina is trouble-free in the landscape and is readily available from mainstream retail nurseries. Seeds germinate easily, and seedlings may need to be controlled around mature specimens. Prune it back when young if you prefer it to be a multistemmed shrub, or seek out the cultivar named 'Compacta' that stays compact and bushy to about 6' tall without pruning.

Although this is positively one of the very best trees to attract butterflies, all members of this genus have poisonous fruits and irritating sap, so take this into consideration if young children will have access to the tree. It may be deciduous in winter.

Bahama Swamp Bush

Pavonia bahamensis
Malvaceae (Mallow Family)

Flowering Season: All year.

Native Range: Bahamas.

Comments: Very few flowering trees attract more hummingbirds in southern Florida than this Bahamian endemic species. It reaches about 14' tall but produces flowers when very young. Green, hibiscus-like, ¾" flowers are produced on long spikes, and you can literally squeeze nectar from the flowers. Not a day goes by when it is not blooming, and hummingbirds vigorously defend the territory around the plant. Orioles, vireos, and warblers also sip the abundant nectar from the blossoms, as do skippers and other butterflies.

Although not commonly cultivated, it is often available from plant sales at Fairchild Tropical Botanic Garden in Coral Gables. It requires ample sunlight throughout the day and is easily propagated by cuttings or air-layers. A mature specimen will be multitrunked, forming a dense thicket of stems like a large shrub.

It is hardy in Zones 10B and 11 and may need winter protection north of this region. A very similar species called mangrove mallow (*Pavonia paludicola*) is native to Florida (Miami-Dade and Monroe County mainland) but is not cultivated. It grows among mangroves that line the Rodgers River and Broad River in Everglades National Park (Monroe County) and around Black Point Marina near Biscayne Bay (Miami-Dade County).

San Martín Sanchezia

Sanchezia sanmartinensis
Acanthaceae (Acanthus Family)

Flowering Season: All year.
Native Range: Peru.
Comments: This relatively small ornamental tree is globally endangered due to habitat destruction in its native Peru. It has only recently found its way into cultivation in the United States, mostly through Internet nurseries, but Fairchild Tropical Botanic Garden in Coral Gables distributes plants occasionally during its annual plant sales. It was named for the San Martín region of northern Peru where it was first discovered.

The leaves are hairy with tubular red flowers emerging at the branch tips from red, hairy bracts that combine to make an interesting and attractive floral display. It cannot be successfully grown outdoors in the cooler sections of Zone 10A. Even in the southernmost counties of Florida, it may still require frost protection in winter.

This species can reach 14' tall and prefers partial shade with rich soil. When in flower it will receive a lot of attention from hummingbirds and some of the larger species of butterflies. Due to its rarity in cultivation, it will also receive deserved attention from visitors to your home garden. It grows easily from cuttings if you care to share it with fellow gardeners.

Desert Senna

Senna polyphylla

Fabaceae (Pea Family)

Synonyms: *Cassia polyphylla.*

Flowering Season: All year.

Native Range: Tropical America.

Comments: The long, arching branches of desert senna produce a pleasant umbrella-like canopy cloaked with bright yellow blossoms off and on throughout the year. The compound leaves comprise many tiny leaflets, and these serve as larval food for the orange-barred sulphur and cloudless sulphur butterflies. The leaflets are so small and numerous that feeding damage from butterfly larvae will never be noticed. Adult butterflies also visit the flowers for nectar.

Desert senna requires little care and is a charming small, shrubby tree that reaches 8–10' in height. It is best used as a freestanding specimen in full sun to show off its brilliant floral display and to maintain its tidy canopy. It could even be used formally on either side of an entranceway or to line a small walkway.

Young plants have a disorganized branching habit, but no corrective pruning is necessary since its shape will regularize naturally once it begins to mature. It is hardy through Zone 10 and is widely grown by nurseries in central and southern Florida. 🦋 🐛

Five Fingers
Tabebuia bahamensis
Bignoniaceae (Trumpet Creeper
Family)

Flowering Season: All year.
Native Range: Bahamas and Cuba.
Comments: Other common names in the Bahamas for this quaint tree are beef-bush, gumwood, above-all, and white-cedar, but the name five fingers is in frequent use wherever it is grown throughout the Caribbean and refers to the five fingerlike leaflets.

The crinkled flowers are trumpet-shaped, to about 1" wide, and are very light pink to nearly white. Hummingbirds are very frequent visitors to the flowers, including the Bahama woodstar and Cuban emerald in its native range, and the ruby-throated hummingbird in tropical Florida. Butterflies visit the flowers throughout the year.

This is not a tree you will easily find in Florida nurseries, but you can find it in specialty nurseries in Miami-Dade County. Five fingers may reach 20' tall or more in Florida gardens, and its narrow, open canopy allows plenty of sun to reach understory plants. It is also a wonderful tree on which to grow epiphytic orchids and small bromeliads.

It is pretty enough to be displayed in front yards and would even make an excellent neighborhood street tree. If you're looking for a pretty tree for a small yard, or to grace the border of a patio close to your home, this should be a top choice.

Yellow Elder

Tecoma stans
Bignoniaceae (Trumpet
Creeper Family)

Flowering Season: Mostly winter and spring.
Native Range: Tropical America.
Comments: Yellow elder is the national flower of the Bahamas and also the official flower of the U.S. Virgin Islands. It is sparingly naturalized in southern Florida, typically in canopy gaps and fringes of hardwood hammocks, but was once championed as a native component of Florida's flora in the early literature. Some Florida nurseries still mistakenly offer it as a Florida native. Through cultivation, it is now widespread in warm regions of the world.

In flower, yellow elder is covered with a profusion of trumpet-shaped, bright yellow flowers. Although it can reach 16' tall, it can be maintained as a large shrub through pruning. Give it a prominent place in your landscape, because it is dazzling when it blooms.

Skippers are the most common butterfly visitors, often crawling down the throat of the flowers to access nectar. Hummingbirds also frequent the flowers along with many other nectar-seeking birds, including the red-whiskered bulbul that is naturalized in Miami-Dade County from Asia. Renowned Florida nurseryman Henry Nehrling regarded yellow elder as the most beautiful of all flowering trees. Cultivars include 'Mayan Gold' and 'Gold Star.'

Wild Lime

Zanthoxylum fagara

Rutaceae (Rue Family)

Flowering Season: All year.

Native Range: Central and southern Florida and from Texas south into tropical America.

Comments: Wild lime is a spiny Florida native tree to 12' tall or more with compound leaves bearing small, rounded, aromatic leaflets. Judicial pruning may be necessary to keep the plant within bounds where space is limited. It could even be maintained in a large container and kept shrubby by pruning.

The clusters of yellow, axillary flowers are fragrant, and the small, shiny, black seeds are consumed by a variety of songbirds. Wild lime is a preferred larval host plant of the giant swallowtail (larva pictured), so it comes highly recommended. The endangered Schaus' swallowtail and Bahamian swallowtail of the Upper Florida Keys also occasionally use it as a host plant.

Its wickedly sharp thorns can be an attribute if you have areas where you do not want foot traffic or trespassers. Another spiny native species to consider is Hercules' club (*Zanthoxylum clava-herculis*), but it is not as readily available as wild lime. Various citrus trees are another option, but they have been omitted from this guide due to the spread of citrus canker and citrus greening, two bacterial diseases that are threatening the entire citrus industry in Florida.

Shrubs

This section includes woody or semiwoody plants that typically do not reach tree stature, attaining average heights of 4 to 12 feet. Shrubs are very versatile in the landscape and therefore merit special consideration.

Lion's ear, *Leonotis leonurus.*

Flowering Maple

Abutilon × *hybridum*
Malvaceae (Mallow Family)

Flowering Season: All year.

Native Range: A hybrid of unknown parentage.

Comments: No, it's not a true maple, but the leaves make it look like one. The genus *Abutilon* ranked second in a "best hummingbird plants" poll taken on an Internet forum (second to *Salvias*), so it is obviously a great hummingbird attractor. This semitropical hybrid is available with red, purple, pink, orange, yellow, white, and even striped flowers. In cold temperate regions, flowering maples are treated as annuals, but in tropical Florida, they may remain evergreen and shrublike, with some reaching 6' tall. The cultivar pictured is 'Louis Sasson,' which tops out at about 4' in height with a bushy shape and 1¼" flowers. As a hummingbird magnet it is highly recommended. Butterflies fuel up on the nectar as well. Its only requirements are ample sunlight and soil that remains moderately moist. If you grow it in a container, ensure that it is watered regularly or it will quickly wilt.

Other selections include 'Apricot Glow,' 'Clementine,' 'Halo,' 'Old Rose Belle,' 'Pompeii,' and 'Sugar Plum.' Look for them through Internet sources, because you will not likely find them in tropical Florida nurseries.

Sweet Almond Bush

Aloysia virgata
Verbenaceae (Verbena Family)

Flowering Season: All year.
Native Range: Tropical America.
Comments: The intensely perfumed flowers of sweet almond bush are the best reason to add this plant to your garden. Bees love the flowers, and butterflies like hairstreaks, crescents, and blues are attracted as well. If you have Atala butterflies, then this is a must for your garden, because they cannot seem to get enough of the nectar. It reaches a height of about 12', with long, brittle branches covered with rough, lance-shaped leaves. The ever-present spikes of flowers fill the air with a fragrance like fine perfume.

Sweet almond bush must have full sun during the heat of day for optimum flowering potential and to maintain a compact shape. It is best used as a free-standing specimen or among plants of lower stature. Pruning will help maintain a more compact canopy.

Other members of this genus are used as a lemon flavoring for French liqueurs and as a fragrant sedative tea. You can dry the flowers of sweet almond bush and use them in potpourri. It is hardy in Zones 10 and 11 but may survive winter temperatures in coastal regions of Zone 9.

Southern Indigobush

Amorpha herbacea
Fabaceae (Pea Family)

Flowering Season: Spring and summer.

Native Range: Along the Coastal Plain from South Carolina to Alabama south through most of Florida.

Comments: This small, colorful, butterfly-attracting native shrub is regrettably little known in cultivation, but there are some specialty nurseries that propagate it for native-plant enthusiasts. In spring and summer, the plant is adorned with 6–8" spikes that are lined with rich blue or white flowers. These attract a variety of butterflies, bees, and other insects that come seeking a nectar feast.

The closely related crenulate leadplant (*Amorpha crenulata*) is endemic to southern Miami-Dade County and differs in having conspicuously crenulate (scalloped) leaflet margins. It is only grown by enthusiasts and is rarely seen in cultivation, even in Miami-Dade County. Both are drought-tolerant once established and provide interesting color and sculpture to any garden setting.

Southern indigobush is small enough to be grown in a container on a sunny patio and typically only reaches 3–4' in height with a slightly wider spread. In addition to being a nectar source, it is also a larval host plant of the southern dogface and silver-spotted skipper within their range. 🌶 🦋 🐛

Hummingbird Bush

Anisacanthus quadrifidus var. *wrightii*
Acanthaceae (Acanthus Family)

Flowering Season: Summer into winter.
Native Range: Texas and northern Mexico.
Comments: The sprightly, 4-lobed, red flowers of hummingbird bush are in abundance from early summer well into winter and are a perennial favorite of hummingbirds and butterflies. Although it is decidedly drought tolerant, it looks more attractive if it receives regular watering in the dry season. In cold temperate regions, it is cut down near the ground to overwinter and then it rebounds in spring, but this is not necessary in tropical Florida. It is typically deciduous, so leaf loss in winter is normal.

In the garden expect it to be shrubby to about 4' tall, although it can reach 6' tall in prime growing conditions. It likes full sun and rich soil but will tolerate light shade and many soil types. Use it along sunny borders of your garden or plant it in sunny gaps between trees. It can also be successfully maintained in a pot for a sunlit porch, deck, or patio.

Hummingbird bush is not overly popular in the Florida nursery trade, but it is worth seeking in nurseries that specialize in flowering plants or via Internet sources.

Hartweg's Aphelandra

Aphelandra hartwegiana
Acanthaceae (Acanthus Family)

Flowering Season: Spring into fall.
Native Range: Colombia and Panama.
Comments: This relative newcomer to horticulture in Florida shows great promise as a colorful addition to gardens in southern Florida. The plant has a somewhat leggy growth habit but makes up for it with a tall spike lined with bright orange bracts topping each stem. One to several bright yellow flowers open in slow succession over a period of many weeks. Each spike can reach more than 2' tall as it progresses. It is easily propagated by cuttings and can be pruned periodically to maintain a more bushy appearance.

Hummingbirds and butterflies, mostly swallowtails and sulphurs, visit the flowers. It performs best in a sunny, protected location. It will require winter protection during hard freezes, but if it is killed to the ground it will resprout from the base in spring.

Specimens can be seen at Fairchild Tropical Botanic Garden in Coral Gables. The acanthus family is one that gardeners in tropical Florida should pay close attention to, because it has many members that attract both hummingbirds and a wide array of butterflies. Some of the best are included in this guide.

Panama Flame

Aphelandra panamensis
Acanthaceae (Acanthus Family)

Flowering Season: Summer into winter.
Native Range: Panama.
Comments: The shocking pink flowers of this Panama native brighten lightly shaded or sunny areas and are a nectar feast for hummingbirds. Swallowtails, sulphurs, and other butterflies find the flowers irresistible as well.

You will likely have difficulty finding this species because it is relatively new to cultivation, but it is available in select nurseries in southeastern Florida. There are about 175 species in the genus *Aphelandra,* and most of them are hummingbird pollinated in their native tropical American countries. Florida gardeners in Zones 10 and 11 should learn more about this genus and acquire plants of other species when the opportunity arises. They all lend a distinct tropical ambience to gardens, and many species are unknown in cultivation, which is surprising because many are simply eye-popping when in flower.

Panama flame reaches 6–8' tall in cultivation; pruning will help alleviate its leggy growth habit. Cuttings root easily and will create more plants for you to grow or share. It will look best in partial shade or, at least, where it has transitional shade during the heat of day.

Hierba del Camaron

Aphelandra schiedeana
Acanthaceae (Acanthus Family)

Flowering Season: All year.
Native Range: Mexico and Central America.
Comments: The Spanish name of this striking shrub means "grass shrimp." It reaches 6–8' tall and prefers light filtered through trees, or full sun in the morning hours and transitional shade through the afternoon. Among the members of the genus *Aphelandra* in cultivation, this species may prove to be one of the most difficult to find, but there are some specialty nurseries in southeastern Florida that offer it. Also check plant sales at special events offered at botanical gardens.

Hummingbirds delight in the flower nectar, as do large species of butterflies. Also watch for sphinx moths visiting the blossoms around twilight as they hover in front of the flowers and sip nectar exactly like hummingbirds. This species flowers off and on throughout the year but most especially during the cooler winter months, which is important for wintering hummingbirds in the southernmost counties. As for cold hardiness, it would be wise to cover it with a sheet or flannel blanket if there is a chance of frost, or worse.

If you choose to prune it to keep it bushy, save the cuttings to give to your favorite gardener friends.

Panama Queen

Aphelandra sinclairiana
Acanthaceae (Acanthus Family)

Flowering Season: Winter and spring.

Native Range: Central America.

Comments: The contrast of pinkish-violet flowers and orange, shell-like bracts makes Panama queen a winner for any garden setting. It prefers bright shade but will grow nicely in areas where it receives full sun in the morning. It can grow to 10' tall, but its long, leggy stems should be pruned occasionally to create an abundance of shorter stems that will flower in the upcoming season.

Hummingbirds will frequent the flowers, as will large species of butterflies. Panama queen is more commonly available than other members of the genus, but you may still have difficulty finding it in mainstream garden centers. Check specialty nurseries or Internet sources. It roots easily from cuttings, if you know a gardener friend who grows it.

In the landscape, Panama queen looks best tucked in among shorter shrubs where it can show off its attractive blossoms above its garden mates. Few other plants offer such outstanding colors in shady situations, so scatter several of these around your garden for splashes of color in winter. If you are short on space, then grow it in a decorative container placed in light shade.

Red Aphelandra

Aphelandra tetragona

Acanthaceae (Acanthus Family)

Flowering Season: Midsummer into winter.

Native Range: Venezuela.

Comments: The dazzling, fanlike flowers of this tropical shrub flare out from branched spikes like eye candy for your garden. Any self-respecting hummingbird will battle over the territory surrounding it, and large butterflies, especially swallowtails, will probe the flowers for nectar.

It is not widely available in the Florida nursery trade, but check botanical garden plant sales or nurseries that specialize in tropical flowering plants. It is readily available through Internet sources if you cannot find it elsewhere. Rest assured that it is worth every effort to add it to your collection.

Red aphelandra is hardy in Zones 10 and 11 but will likely require winter cold protection in the northern sections of Zone 10A. Like other members of this genus, it tends to be somewhat leggy, but this can be corrected by selective pruning to encourage a bushier habit. Pruning also increases flowering potential, because the flower spikes emerge from the branch tips.

It tolerates full sun but will look and perform best in the partial shade provided by palms or light-canopied trees. Water it moderately during the dry season. Pruning in late winter will encourage new spring growth.

Panama Rose
Arachnothryx leucophylla
Rubiaceae (Madder Family)

Synonyms: *Rondeletia leucophylla*.
Flowering Season: All year, but especially spring through summer.
Native Range: Mexico to Panama.
Comments: Panama rose deserves a standing ovation because there is hardly a more beautiful sight than watching butterflies and hummingbirds dart from one cluster of flowers to another all day long. Give it a prominent place in your garden where you can sit and watch all of the activity. It thrives in full sun and is among the most colorful of shrubs, reaching a height of about 8' at maturity.

It is hardy into Zone 9, so there is no need for winter cold protection anywhere in tropical Florida. Once established, it requires little care and rarely needs pruning unless you wish to keep it smaller. It would make an outstanding tall privacy hedge to add year-round color along your property line.

Various species of butterflies visit the flowers, and from fall to spring it is a favorite of ruby-throated hummingbirds in Zones 10 and 11. The flower clusters are about 2" across, and individual flowers are uniformly bright pink. It is available from nurseries that specialize in flowering plants or from Internet sources. 🐦 🦋

Sea Lavender

Argusia gnaphalodes
Boraginaceae (Borage Family)

Flowering Season: All year.

Native Range: Central and southern Florida, West Indies, and tropical America.

Comments: Lucky you, if you live in a beachfront home—but don't fret if you live inland, because sea lavender will perform nicely away from its beach dune habitat if its horticultural needs are met. This species is critically rare due to habitat destruction from oceanfront development, so for those of you who live in homes where dunes once existed, here's your chance to repent your sins by planting this wonderfully attractive, state-listed endangered species.

It forms dense mounds of stems to about 4' tall and 6' wide with the leaves crowded toward the branch tips. The leaves are silvery gray green and contrast nicely with other plants. The butterfly-attracting flowers are white, turning pink as they age, and are present all year.

In inland locations, sea lavender will do best if planted on a mound of gravelly sand to imitate a beach, or you can grow it in a large container filled with commercial cactus soil amended with perlite or coarse gravel. It will not fare well if planted in heavy soils that do not drain quickly. One other requirement is full sun all day long or at least through midday. Check nurseries that specialize in Florida natives.

Pineland Strongback

Bourreria cassinifolia
Boraginaceae (Borage Family)

Synonyms: *Bourreria havanensis.*
Flowering Season: All year.
Native Range: Miami-Dade County, Florida Keys, Bahamas, and West Indies.
Comments: Southern Florida gardeners need to find a prominent place in a sunny location for this quaint, state-listed endangered species. Although it is endangered in the pine rocklands of Miami-Dade County and on Big Pine Key in the Lower Florida Keys, it is gaining popularity as a landscape plant in the southernmost counties.

Pineland strongback can reach small-tree stature but is typically a rounded shrub to about 8' tall with a dense canopy. The leaves are much smaller than the related Bahama strongback (*Bourreria succulenta*) and rough strongback (*B. radula*), but the flowers of all three species are equal in their appeal to butterflies and hummingbirds. The flowers are especially attractive to Julias, Gulf fritillaries, zebra longwings, large species of sulphurs, a variety of skippers, and even sphinx moths.

Pineland strongback is useful as a barrier screen, included in a group planting with other shrubs, or grown in a large container for a patio.

The name *strongback* is often corrupted as "strongbark" in Florida, but the name originated in the Bahamas, where it is used in bush medicine to give men a "strong back," with lovemaking implications. Songbirds relish the small orange fruits.

Hummingbird Flower

Bouvardia ternifolia
Rubiaceae (Madder Family)

Flowering Season: All year.

Native Range: Texas, Arizona, New Mexico, and Mexico.

Comments: With a name like hummingbird flower, how can any respectable hummingbird resist? Indeed, every hummingbird within eyesight of this plant races over to gorge on nectar and claim territorial rights to the plant. While one chases an intruder away, another sneaks in for a quick helping of its own.

Hummingbird flower is not well-known to Florida gardeners, but it is a prized native perennial for hummingbird and butterfly gardeners in the American Southwest. It prefers acidic, well-drained soil and may become chlorotic in the alkaline soils of South Florida, but this can be remedied with applications of chelated iron. Another option is to maintain it in a pot filled with a peat-based potting mix.

This species will reach 3' tall at maturity with clusters of red-orange flowers at the branch tips practically every day of the year. It is very heat tolerant, so give it full sun for best flowering potential, and place it where you can enjoy the parade of butterflies and hummingbirds around the blossoms.

Summer Lilac

Buddleja davidii

Scrophulariaceae (Figwort Family)

Flowering Season: All year.

Native Range: China.

Comments: Summer lilac is hardy to Zone 5 and generally prefers cooler, drier summers than what is provided in tropical Florida. It is a woody shrub, typically from 4–6' tall, but can reach twice those heights. The leaves are green above and white beneath, with showy, terminal spikes of flowers. The typical variety bears orange-centered lilac flowers, but many varieties and cultivars are available, with colors including pink, mauve, dark purple, and white. The flowers are delectably fragrant, although the scent is imperceptible to some people.

In Zones 10 and 11, the white-flowered form (pictured) can best handle our summer heat and elevated humidity. Many southern Florida gardeners simply grow the other color forms as annuals, expecting them to perish when summer takes its toll. Hummingbirds visit the flowers; this is also a superior nectar source for a wide range of butterflies.

Department store garden centers may periodically offer summer lilac, or check nurseries that specialize in flowering plants. Propagation is either by seed or tip cuttings. This plant is invasive in some parts of the United States, but not in Florida. Also consider weeping butterfly bush (*Buddleja lindleyana*).

Madagascar Butterfly Bush

Buddleja madagascariensis
Scrophulariaceae (Figwort Family)

Flowering Season: Winter and early spring.

Native Range: Madagascar.

Comments: This bold plant is much larger than the previous species and has a mounding growth habit, with long, sprawling stems that may reach more than 10' in height with an even greater spread. It produces attractive and abundant clusters of yellow flowers with a delicate perfume. The flowers fade to white with age. Specialty nurseries that grow butterfly-attracting plants in southern Florida occasionally offer this species, and it is a very worthy addition to tropical Florida gardens if you have the space it requires. Judicial removal of rooted stems will be necessary to keep it within its bounds.

Butterflies literally swarm around a flowering specimen; the flowers are especially attractive to monarchs, queens, white peacocks, and red admirals. Another virtue is that it flowers through the winter months when other plants are long past their blooming season.

Propagation is by cuttings. It requires full sun and seasonal pruning to maintain any semblance of a compact growth habit. If this species sounds too rambunctious for you, then there is a yellow-flowered, better-behaved hybrid called *Buddleja* × *weyeriana* 'Honeycomb' that may fit the bill.

Fairy Duster
Calliandra tergemina var. *emarginata*
Fabaceae (Pea Family)

Synonyms: *Calliandra emarginata.*
Flowering Season: All year.
Native Range: Southern Mexico to Panama.
Comments: Fairy duster can best be described as a dwarf version of the more common red powderpuff (*Calliandra haematocephala*) treated previously in this guide. Although the flowers are somewhat short-lived, they appear continuously, offering year-round color, and they are a favorite of ruby-throated hummingbirds as well as butterflies, especially ruddy daggerwings, swallowtails, and large species of sulphurs.

Fairy duster requires ample water with well-drained soil, or it will become anemic and stunted, so grow it where it can be watered. It may reach about 8' tall with a spread to about 5' wide. For best flowering, plant it where it will receive strong sunlight during much of the day, and to avoid nitrogen and iron deficiencies, keep it on a monthly fertilizer regimen. Through judicial pruning it can be kept in a container, if that suits your situation better.

With their conspicuous stamens, the flowers resemble small powder puffs and are 2" wide. Fairy duster is hardy in Zones 10 and 11 and should be a welcomed addition to your garden.

American Beautyberry

Callicarpa americana
Lamiaceae (Mint Family)

Flowering Season: Spring into fall.

Native Range: Southeastern United States and the West Indies.

Comments: American beautyberry is a popular shrub among gardeners throughout Florida and is readily available in the nursery trade. Tight, rounded clusters of pink flowers appear along the stem in the leaf axils and are followed by round, very ornamental clusters of bright violet fruits. There is also an albino form in cultivation with white flowers and fruits. Butterflies visit the flowers, and birds like mockingbirds, gray catbirds, and American robins feast heartily on the fruits.

The leaves of this multistemmed shrub are highly aromatic when crushed, and a mature specimen may reach 6' in height with an equal spread. To maintain it as a smaller, more compact plant, cut the stems down to the ground every two or three years and allow it to sprout new stems. This may sound radical to some gardeners, but in the wild, it is kept shrubby when fire burns it to the ground.

American beautyberry looks especially nice when planted with saw palmetto (*Serenoa repens*), which shares its native habitat. It is exceptionally drought tolerant and is relatively trouble-free. Taxonomists recently moved this genus from the verbena family (Verbenaceae) into the mint family (Lamiaceae).

Boar Hog Bush

Callicarpa hitchcockii
Lamiaceae (Mint Family)

Flowering Season: Spring to fall.

Native Range: Bahamas and islands off of northern Cuba.

Comments: In the Bahamas, this species grows in pinelands, coppices, and open savannas. It is not known to be cultivated in the United States except by botanical gardens, namely Fairchild Tropical Botanic Garden in Coral Gables, but it deserves horticultural attention. It is closely related to the previous species but differs by its growth habit and stiff, narrow leaves. It is a multibranched, 4–6' shrub with stems that can sometimes seem vinelike. The clusters of small, pink flowers are very attractive to butterflies. The ¼" fruits are royal purple to bluish, and it is very ornamental when flowering or fruiting.

Boar hog bush requires full sun and well-drained soil for optimum growth. It can be grown in a container but will need pruning to keep it from becoming rangy. Sulphurs seem to be especially attracted to the flowers as well as species of skippers.

This will be a difficult plant to find unless it is offered at botanical garden plant sales. If you decide to fly to the Bahamas to gather your own seeds, look for it on Eleuthera, New Providence, or Andros. You can thank me later.

Giant Milkweed

Calotropis gigantea
Apocynaceae (Dogbane Family)

Flowering Season: All year.
Native Range: Tropical Asia.
Comments: Butterfly gardeners have made giant milkweed popular because the leaves serve as larval food for queen butterflies and occasionally for monarchs. One advantage of giant milkweed as a larval host plant is its large leaves. One butterfly larva will only eat a portion of a single leaf before maturing enough to pupate, a distinct advantage when trying to maintain populations of queen butterflies. A few larvae can entirely strip the leaves, flowers, and stems off of members of the related genus *Asclepias*. Still, you will definitely notice feeding damage on the leaves by butterfly larvae, so do not use giant milkweed as a centerpiece if chewed leaves bother you.

Cut stems root easily, and periodic pruning will help maintain a more orderly shape. Like other milkweeds, it attracts milkweed bugs and aphids, but these can be deterred with strong blasts from the garden hose or soapy water. It requires full sun and well-drained soil for optimum growth and is very drought tolerant. The distinctive bluish-gray leaves and oddly attractive starlike flowers lend an interesting appeal to any landscape design. The sap contains steroidal heart toxins that can be fatal if eaten. 🐛 ☠

Buttonbush

Cephalanthus occidentalis
Rubiaceae (Madder Family)

Flowering Season: Spring and early summer.
Native Range: North America and Cuba.
Comments: This Florida native freshwater wetland shrub produces sweetly fragrant ornamental flowers that attract a variety of butterflies as well as bumblebees and other nectar-seeking insects. It requires permanently wet soil; if you do not have a pond, lake, stream, or water garden, it can be grown successfully in a large pot partly submerged in water. The only horticultural attention it requires is periodic pruning to maintain a compact growth habit and to encourage flowering each season. Even without pruning, the terminal bud aborts, creating a forked branching habit. In cold temperate regions it is deciduous, but it is evergreen in southern Florida unless there is a particularly dry, cold winter.

Although buttonbush can reach small-tree stature, it is typically a shrub 4–8' tall. The tight clusters of flowers resemble round pincushions, and its ease of culture and reliable flowering are genuine attributes. It is generally available only from nurseries that specialize in native plants or from water garden nurseries. It is decidedly cold tolerant as demonstrated by its natural range, which extends into Canada, and it is common in wetlands throughout mainland Florida. Propagation is by seeds, cuttings, or air-layers.

Jack-in-the-Bush

Chromolaena odorata
Asteraceae (Aster Family)

Flowering Season: All year.

Native Range: Central and southern Florida to tropical America.

Comments: Pinkish-violet heads of small flowers engulf this native shrub, and butterflies will literally swarm around a flowering specimen. This species is seldom, if ever, propagated for sale, even by nurseries that specialize in native plants, but it grows easily from seeds or cuttings. It is a worthwhile addition to your garden because it is such a preeminent butterfly attractor when in bloom. In the wild, it is typically found along the margins of hardwood forests or in sunny canopy gaps, but it can also be found colonizing disturbed sites, such as overgrown lots or roadsides.

It requires full sun or strong light during most of the day for best flowering potential. It is very fast growing; it will flower in its first year from seed, and the seedlings may appear on their own anywhere there is bare soil. Plant it in the landscape along the sunny edges of group plantings just as it appears in nature. It cannot be stressed enough that this is a serious butterfly magnet.

Pagoda Flower

Clerodendrum paniculatum
Lamiaceae (Mint Family)

Flowering Season: Late spring through early winter.
Native Range: Southeast Asia.
Comments: A well-grown stand of pagoda flower is a stunning sight in the landscape when each 6–8' stem is topped by a mass of small, red flowers arranged in an airy, pyramidal panicle. There is a white-flowered form as well, but it is much less commonly seen. A single panicle can reach over 2' tall and have several hundred blossoms open at once. A horticultural concern is that the plant spreads from root suckers and can quickly claim a large area if left untended. Root suckers can be dug up, potted, and given to others, or they can simply be cut down to the ground as a gardening chore each year.

Pagoda flower attracts an array of butterflies, especially sulphurs and swallowtails, as well as hummingbirds. It requires strong light and open space to spread. It is available in specialty nurseries, but if you find fellow gardeners who cultivate it, they will likely be more than happy to rid themselves of a few root suckers. It sets seed only sparingly and is not naturalized in Florida like its close relative, Java glory bower (*Clerodendrum speciosissimum*), which is not recommended due to its aggressive, weedy tendencies from seed.

Chains of Glory

Clerodendrum schmidtii
Lamiaceae (Mint Family)

Flowering Season: Winter.

Native Range: Thailand, Laos, and Cambodia.

Comments: No words can describe the beauty of a flowering *Clerodendrum schmidtii*. Inch-wide, fragrant, pure white flowers appear on branched, cascading, red inflorescences to about 18" long, and they time their appearance right around Christmastime. The unopened flower buds resemble white raindrops, pearls, or even light bulbs, so the plant exudes charm even before the blossoms open.

It is rare in cultivation and seen mostly in botanical gardens, but specialty nurseries in southeastern Florida offer it. If all else fails, it is available through Internet sources, but also check botanical garden plant sales.

It is shrubby to about 6' in Florida but reaches small-tree stature in its native range, where it grows in hot, humid jungles. Toss a flannel blanket over it if temperatures are supposed to dip into the 40s or below. Bring it indoors if it is in a pot.

It looks much like the commonly cultivated *Clerodendrum wallichii* but is much classier; it's like comparing a fourteen-carat diamond to the Hope Diamond. Go out of your way to find it, and the hummingbirds and butterflies will simply be added bonuses.

Thailand Powderpuff

Combretum constrictum
Combretaceae (Combretum Family)

Flowering Season: Sporadically all year.
Native Range: Tropical Africa.
Comments: This species can be treated either as a shrub or a vine. Planted on its own, it forms a large shrub with long branches that bear sharp, hooked thorns. The thorns are used for climbing high into trees if given the opportunity, but this should be avoided unless you have a tree large enough to support it. The branches are easily capable of reaching the top of a 20' tree.

Round clusters of 1½–2" red flowers with long stamens are produced periodically throughout the year and are visited by butterflies and hummingbirds. It is very attractive as a freestanding specimen but will require pruning to keep it manageable. It would benefit by having something to support its branches, such as a freestanding section of fence or a strong trellis. It could also be trained to climb up an arbor, but the flowers would not be as visible once it reached the top.

Remember the sharp thorns when selecting a place for it in your garden, and keep your limb loppers handy. It is available from specialty nurseries and from Internet sources.

Bahama Manjack
Cordia bahamensis
Boraginaceae (Borage Family)

Flowering Season: All year.

Native Range: South Florida (Miami-Dade County), Bahamas, and Cuba.

Comments: Bahama manjack is a little-known shrub grown by native-plant hobbyists in the southernmost counties of Florida. It was first discovered growing wild in Florida in a remnant pineland east of Everglades National Park in Florida City, but that site was destroyed for development. It is not known elsewhere as a wild plant in Florida, but because of its historic occurrence it is a state-listed endangered species. It was introduced into the Florida nursery trade from seeds collected by the author on the island of Abaco in the Bahamas and is still today offered by a select few nurseries in Miami-Dade County.

Because of its upright stature, it will take up less room than butterfly sage (treated next) and is very handsome with its shiny leaves, small white flowers, and brilliant red fruits that are savored by birds. The flowers are visited by a virtual parade of butterflies.

As for cold hardiness, it may require some degree of cold protection in the cooler sections of Zone 10A. Give Bahama manjack full sun for a nice compact canopy. It comes highly recommended as an addition to butterfly gardens in tropical Florida.

Butterfly Sage

Cordia bullata var. *globosa*
Boraginaceae (Borage Family)

Synonyms: *Cordia globosa.*
Flowering Season: All year.
Native Range: Southern Florida and tropical America.
Comments: Although the natural range of this species in Florida is restricted to Miami-Dade and Monroe counties, it can be successfully grown northward into coastal Central Florida. Elsewhere it will require winter cold protection.

It forms a rounded shrub 4–8' tall with small, cupped, white flowers that butterflies simply cannot resist. Practically every species of butterfly in your neighborhood will be at the flowers, from large swallowtails to tiny hairstreaks and blues. Ruddy daggerwings especially favor the flowers. As a bonus, northern mockingbirds and gray catbirds feast on the fleshy red fruits, spreading seeds as "gifts" to your neighbors. It is a definite must for every resident in tropical Florida who is serious about attracting butterflies to their garden.

Butterfly sage looks best when included in a group planting in full sun, but give it plenty of room because it spreads to about 6' wide or more. It can only be found in nurseries that specialize in Florida native plants, especially in the southernmost counties, but is well worth going out of your way to acquire for your garden.

Azulejo

Cornutia pyramidata
Lamiaceae (Mint Family)

Synonyms: *Cornutia grandifolia.*
Flowering Season: Spring into fall.
Native Range: Mexico into tropical South America.
Comments: The large fuzzy leaves combined with branched spikes of bright blue flowers make this little-known shrub a pleasing addition to tropical Florida gardens. The 4-lobed flowers that appear from fuzzy buds over a period of many months are alluring to every butterfly within eyesight. The leaves have a strong odor when rubbed, but it is not unpleasant. It grows easily from cuttings and may root wherever branches touch the ground. This should be discouraged so the plant does not outgrow its bounds. Although seldom found in nurseries, check plant sales at Fairchild Tropical Botanic Garden, Butterfly World, or some of the specialty nurseries in southern Florida.

Azulejo (Spanish for "bluebird") can reach heights of 10' or more with an equal spread, but you can prune it to keep it smaller as a more compact shrub. It grows rapidly and has brittle branches that snap off easily, so plant it in a protected location. It will flower in a considerable amount of shade but will take on a more attractive shape if given full sun. Some botanists place this genus in the verbena family (Verbenaceae).

Orange Cuphea

Cuphea schumannii
Lythraceae (Loosestrife Family)

Flowering Season: All year.
Native Range: Mexico.
Comments: To say hummingbirds find the flowers of this shrubby species attractive would be a major understatement. At first light of day, hummingbirds fly straight to the tubular orange blossoms and then return over and over for more helpings of sweet nectar throughout the day. Orange cuphea must be planted in soil that drains well, and it must have strong sunlight through most of the day. As it typically reaches only about 4' tall, keep it unobstructed by other shrubs to ensure hummingbirds have access to the blossoms. The branches are somewhat vinelike, so if you decide to keep it in a container, it will be best to give it a small trellis to help keep it upright.

Orange cuphea is surprisingly uncommon in Florida gardens despite its pleasing flowers, ease of culture, and its attractiveness to every passing hummingbird. There are several other *Cuphea* species in cultivation, but this one is undoubtedly the best for hummingbirds (see also bat face, *Cuphea llavea,* in this guide). Orange cuphea is cold hardy, so no winter protection is needed in Zones 10 or 11. As an added bonus, large species of butterflies sip nectar from the flowers as well. It's a prizewinner for any garden.

Golden Dewdrop

Duranta erecta

Verbenaceae (Verbena Family)

Synonyms: *Duranta repens.*

Flowering Season: All year.

Native Range: West Indies.

Comments: Golden dewdrop is undoubtedly one of the premier butterfly-attracting shrubs for tropical Florida, but consideration should be given to the poisonous fruits. The round, golden-yellow fruits form attractive clusters, so they may be enticing to young children. They have been known to kill children, dogs, and cats, although birds can eat them. Showy, ½", blue to purple (or white) flowers are in terminal clusters, and a flowering specimen seems to always be surrounded by nectar-seeking butterflies and occasional hummingbirds.

Golden dewdrop requires open space and full sun to maintain its rounded shape. Although it is capable of reaching small-tree stature to 18' in height, it typically matures at about 8' tall with an equal spread, forming a multitrunked shrub.

It can be used as a freestanding specimen or as a tall, dense privacy hedge. Alternating blue- and white-flowered color forms will add interest to a hedgerow, but plant them on 10' centers to avoid crowding. A form with variegated leaves, and an attractive cultivar with ¾" flowers called 'Grandiflora,' are available in the nursery trade. Golden dewdrop is hardy in Zones 10 and 11 but will succeed in warmer parts of Zone 9.

Blue Sage

Eranthemum pulchellum
Acanthaceae (Acanthus Family)

Flowering Season: All year.
Native Range: India and China.
Comments: When British botanist Henry Charles Andrews named this plant in 1800, he gave it the specific epithet *pulchellum,* Latin for "beautiful." And it is no wonder, because when this exquisite shrub is amassed with flowers it is undeniably beautiful.

It reaches about 5' tall and 8' wide in good light and performs best if it receives full sun, with a respite in mid to late afternoon. An open spot on the east side of trees is perfect. Butterflies will be all over the flowers throughout the day along with occasional hummingbirds, so a choice garden location would be near where you like to spend time relaxing.

It looks especially appealing planted in masses, but use it however it works best in your personal garden setting. If you are pressed for space, then grow it in a large decorative pot and prune it when necessary. It will not be difficult to locate in nurseries and garden centers and is trouble-free once it is fully established. Regular watering and a light fertilizer application will get it off to a good start.

Jamaican Poinsettia

Euphorbia punicea
Euphorbiaceae (Spurge Family)

Flowering Season: Principally summer through winter in Florida.

Native Range: Jamaica.

Comments: Although Jamaican poinsettia can reach tree heights in its native Jamaica, it typically tops out at about 10' in Florida gardens. It is exceptionally ornamental with showy, bright red bracts nestled among the leaves, which are bunched near the branch tips. Almost hidden among the red bracts are small yellow flowers that butterflies, especially zebra longwings and Julias, find enticing.

Because of its engaging good looks and its interesting growth character, it deserves a prominent place as a centerpiece in your home landscape. Less formally it lends itself well as a colorful backdrop for shorter plants, but be certain they do not conceal its exquisite branching habit. This can easily be accomplished by using groundcovers, a bed of bromeliads, or low-growing *Salvia* species or pentas. Some plants have pale bracts, so purchase yours when it is flowering so you can ensure it has good color.

It will benefit from a regular fertilizer regimen because it tends to be a heavy feeder. Seeds are rarely produced, and cuttings are difficult to root, so propagation is by air-layers. The common name reflects its alliance with the Christmas poinsettia. The sap can cause skin irritations and this plant is toxic if ingested.

Crossberry

Grewia occidentalis
Malvaceae (Mallow Family)

Flowering Season: All year.
Native Range: Africa.
Comments: Crossberry is a lovely shrub that fully deserves more horticultural attention. It is trouble-free in the landscape, it tolerates full sun or partially shaded situations, it produces attractive blossoms continuously through the year, and the 4-lobed fruits that give it its common name are edible. And, to add to its multitasking résumé, the flowers attract nectar-seeking butterflies.

To locate this species for your garden, you may have to resort to Internet sources if you cannot find it in local nurseries that specialize in tropical flowering plants. Also check plant sales at botanical gardens, especially those in South Florida sponsored by the Tropical Flowering Tree Society.

Crossberry often produces branches at odd angles that may require corrective pruning, but this is of no real horticultural concern. It is moderately fast growing and typically forms a rounded mound of stems to about 6' tall, but it can grow taller. It is tolerant of light frost in Zones 10 and 11 but will be killed to the ground by hard freezes in cold temperate regions. 🦋

Parakeet Flower

Heliconia psittacorum
Heliconiaceae (Heliconia Family)

Flowering Season: Mostly summer through fall.
Native Range: Amazon rainforests.
Comments: Hummingbirds are the principal, and in some cases exclusive, pollinators of many *Heliconia* species in the Neotropics. The unique flower shapes of various *Heliconia* species and the curved bills of their particular hummingbird pollinators coevolved over time for a perfect fit, suiting the needs of both plant and bird. Of the many *Heliconia* species cultivated in Florida, parakeet flower is one of the few that is visited by ruby-throated hummingbirds; most other *Heliconia* species evolved to be pollinated only by specific tropical hummingbirds with curved beaks.

Parakeet flower is commonly sold in mainstream garden centers, and there are dozens of named cultivars with widely varying floral colors from which to choose. The typical color form is pictured.

Because it occurs naturally in Amazon rainforests, it prefers rich, loamy soil, partial shade, and regular watering in the dry season. There are few plants that create a tropical feel in your garden better than *Heliconia* species, but be careful because you can easily become addicted to them. If you do get addicted, there is an International Heliconia Society with a chapter in South Florida where you can get support from other addicts.

Scorpiontail

Heliotropium angiospermum
Boraginaceae (Borage Family)

Flowering Season: All year.

Native Range: Southern United States and tropical America.

Comments: An advisory is necessary for this Florida native species, because it is weedy in garden settings and may sprout from seed far away from its original planting site. Needless to say, gardening chores will be required unless you do not mind it running rampant. It spreads so readily your neighbors two doors down may wonder where it came from.

Blues, crescents, hairstreaks, and other small butterflies frequent the small white flowers, as do monarch and queen butterflies. My recommendation is to overlook its weedy tendencies, because it supplies important alkaloids necessary for male butterflies to attract mates with pheromones.

It can reach about 5' tall with spikes of flowers in 2 parallel ranks that curl under at the tips. In the wild it is often found along the fringes of mangrove forests and saltmarshes, aptly demonstrating its salt tolerance.

Scorpiontail will take on a bushy, compact growth habit in full sun and requires little care. It blooms throughout the year and is quite ornamental. It is hardy from Zone 9 southward and is available from nurseries that specialize in native plants. 🌺 🦋

Garden Heliotrope

Heliotropium arborescens
Boraginaceae (Borage Family)

Flowering Season: Sporadically all year.

Native Range: Peru.

Comments: North of Zone 10, this old-time garden favorite is grown as a summer annual, but in the warmer regions of Florida it is a perennial that may reach heights of 5' or more. The cultivar pictured is 'Blue Wonder,' which has darker flowers than the typical lilac-flowered species. In your garden it will show off its intensely fragrant blossoms off and on throughout the year and will attract every butterfly and hummingbird in the neighborhood each time it blooms.

Because it prefers a break from Florida's summer sun, in the ground it would do well planted on the east side of a tree, so it receives dappled light through the afternoon. It can also be grown in a pot or even a hanging basket while it is young.

'Blue Wonder' is widely available in Florida nurseries, but there are other cultivars with white, pink, mauve, and dark purple blooms if you find those more appealing. The name *heliotrope* means "turning toward the sun," from the mistaken belief that the flowers always face the sun.

Chinese Hat Plant

Holmskioldia sanguinea
Lamiaceae (Mint Family)

Flowering Season: All year.

Native Range: Asia and East Indies.

Comments: Before considering this species for your garden, you first need to mull over how much room you have to donate to it. If left on its own in the open, it can reach 8' tall with a spread three times as wide. Pruning it back after it has finished flowering in spring is the only way to keep it in bounds.

Meanwhile, be aware that you can hardly find a better hummingbird attractor. Butterflies, too. The typical form has curved, tubular, orange flowers subtended by orange, saucer-shaped, persistent calyces. There are cultivars with red, violet, yellow, and green flowers and calyces, but to draw hummingbirds, stick with the orange or red forms.

It can be allowed to climb into trees, but this may be asking for trouble, both for you and the tree, because it will be difficult for you to control and can engulf the entire tree canopy. Given the proper space and pruning, this sprawling shrub is a prizewinner for any garden in tropical Florida, especially for hummingbirds.

Violet Churcu

Iochroma cyaneum

Solanaceae (Nightshade Family)

Flowering Season: All year.

Native Range: Ecuador.

Comments: Members of the genus *Iochroma* are related to such edible plants as tomato, potato, eggplant, and peppers but are also kin to such notoriously poisonous plants as angel's trumpet, devil's trumpet, and belladonna. *Iochroma* falls into the latter group, so keep this in mind if you have young children—all parts are poisonous if ingested.

The flowers range from dark purple to violet, deep blue, red, and pinkish purple, depending on the variety. Hummingbirds probe their beaks into each flower as they fly from one cluster to the next. Large butterflies, especially swallowtails, also take nectar from the long, tubular blossoms.

Violet churcu is a bushy, 6' shrub and should be pruned in early spring to maximize flowering potential. Its gardening needs are average, but it does benefit from a root drench of diluted liquid fertilizer in spring and again in mid to late summer. Look for various color forms and closely related species from Internet sources, or check local specialty nurseries.

Gold Finger

Juanulloa aurantiaca
Solanaceae (Nightshade Family)

Flowering Season: All year.

Native Range: Mexico.

Comments: Although this everblooming, vinelike shrub is typically epiphytic (growing on trees) in its native habitat in Mexico, it is perfectly at home planted in well-drained soil in tropical Florida. In the ground it can reach 8' tall or more with a wider spread, so give it room to ramble. It prefers to grow beneath tall trees in dappled sunlight but will perform well on the north or east side of group plantings where it receives bright light.

Hummingbirds are quite fond of the waxy, orange, tubular flowers that protrude from lighter orange floral bracts, and they dart from one flower to the next until they've had their fill of nectar. The flowers are about 2" long but are partly hidden by the bracts.

Gold finger can be procured from South Florida specialty nurseries, botanical garden plant sales, or Internet sources. If you know someone who grows it, it can be easily propagated by air-layers or from stems rooted in the ground. It is not known if it can be successfully grown as an epiphyte in Florida, but it may be worth trying.

Shrimp Plant

Justicia brandegeana
Acanthaceae (Acanthus Family)

Synonyms: *Beloperone guttata.*
Flowering Season: All year.
Native Range: Mexico.
Comments: Shrimp plant has made a grand comeback in the nursery trade due to the introduction of new, colorful cultivars. The typical form has pale reddish bracts with white flowers, but a popular new cultivar sports brilliant pinkish-red bracts with pink-lipped flowers (pictured), and another handsome form has darker bracts with red flowers. Another popular form, with lime-colored bracts and pinkish flowers, is called 'Fruit Cocktail.'

The flowers are a reliable favorite of hummingbirds and butterflies alike. It will spread on its own by rhizomes and by rooted stems, eventually forming a mass of color. It should be planted in full sun, where it will remain about 5' tall.

It is best used to fill in sunny gaps between trees and taller shrubs, but it can also be planted against a foundation or even in an open bed on its own. It will offer color throughout the year and is trouble-free except for periodic thinning to keep it within its bounds.

The common name refers to the floral bracts, which somewhat resemble the body of a shrimp. It is available in mainstream garden centers and comes with the highest recommendation possible. 🐦 🦋

Pink Candy

Justicia brasiliana
Acanthaceae (Acanthus Family)

Synonyms: *Dianthera nodosa.*
Flowering Season: All year.
Native Range: Brazil, Argentina, Paraguay, and Uruguay.
Comments: The flowers of this tropical species can be pink or red, and both forms are exceptionally pretty. Unlike many members of the acanthus family, it prefers filtered light from tree canopies. Once it matures, it will become the center of attention for any passing hummingbird or butterfly.

Pink candy reaches 2–3' tall with an equal width and looks best when planted in groups to create a bold splash of color. It performs nicely if it receives direct sun in the morning and then partial shade during midday to avoid wilting. The east side of a tree having an airy canopy would be ideal. It is small enough to be kept in a container but will require watering every few days. A light application of fertilizer in the spring will ensure vigorous growth through summer without any nutrient deficiencies.

It is reportedly hardy into Zone 9B, so it will not require winter cold protection in Zones 10 or 11. Nurseries that specialize in flowering plants may grow this species, but if all else fails you will find it to be widely available from mail-order sources on the Internet.

Flamingo Flower

Justicia carnea
Acanthaceae (Acanthus Family)

Flowering Season: Spring to early winter.
Native Range: South America.
Comments: Crowded clusters of flamingo-pink, softly hairy, yawning flowers adorn this elegant shrub from springtime through early winter, and it attracts a parade of butterflies and hummingbirds throughout the day. Because it prefers slightly acidic soil, it succeeds better in Central Florida and northward than in extreme southern Florida. Gardeners in southern Florida should take note of this and grow it in raised beds or containers.

You can thank enterprising nursery growers for coming up with the stunning cultivar 'Radiant,' which sports deep rose-pink blossoms and bronze-green foliage. There is also a white-flowered selection available that looks pretty mixed in with the pink-flowered forms.

Flamingo flower is a forest understory plant in its native habitat, so in garden settings it performs best beneath the canopy of large trees, either in the ground or in raised beds. It tends to grow no more than about 4' tall, so it is suitable for a decorative container or as a colorful border plant in a lightly shaded area.

It should be easy to locate in Florida nurseries, but you can also check plant sales at botanical gardens. Other common names are pink justicia and Brazilian plume.

Summer Sun Justicia

Justicia corumbensis

Acanthaceae (Acanthus Family)

Synonyms: *Justicia leonardii.*

Flowering Season: All year.

Native Range: Mexico.

Comments: Gardeners should pay close attention to all members of the genus *Justicia* because it includes some of the best shrubs to attract both hummingbirds and butterflies. There are quite a number of species that have made it into cultivation, to the delight of gardeners in the tropics, subtropics, and warm temperate regions, including Florida. Let's all hope there are more on the way.

This quaint species is a small, freely blooming shrub with open clusters of tubular, orange flowers. It averages 3–4' tall with a slightly wider spread, so try tucking it in among other shrubs of similar size, or use it to line a sunny walkway, patio, foundation, or anywhere you have full sun. Hard freezes will kill it to the ground, but it will send out new growth in spring. It blooms for months at a time, so it is a worthy subject for any garden. This species is not commonly seen in Florida retail nurseries or garden centers. If you have problems locating it, try searching the Internet for mail-order sources.

Hummingbirds and a wide variety of butterflies visit the flowers.

Mexican Honeysuckle
Justicia spicigera
Acanthaceae (Acanthus Family)

Flowering Season: Sporadically all year; in tropical Florida, throughout summer and fall.

Native Range: Mexico.

Comments: Wow! This shrub is positively sensational when in bloom, and it seduces hummingbirds and many species of butterflies. In addition to its attention-grabbing floral display, it is easy to grow and is tolerant of a range of soil types. It reaches 6' in height with an equal spread, so choose a prominent space for it. You will want to be able to sit back and admire the colorful blossoms and the hummingbirds. The first hummingbird to find it will defend the territory around it, and their territorial antics are quite entertaining. The butterfly that frequents the flowers the most is the giant swallowtail, but some of the large species of sulphur butterflies favor them as well.

Mexican honeysuckle is surprisingly difficult to find in local nurseries, but it is widely available through Internet sources. Local nurseries that specialize in flowering plants are worth checking.

It has a moderately fast growth rate and blooms at a young age. Give it well-drained soil, occasional supplemental watering in the dry season, and a light application of fertilizer in the spring to create a large, healthy, flowering shrub. Propagation is by cuttings.

Parasol Flower

Karomia tettensis
Lamiaceae (Mint Family)

Synonyms: *Holmskioldia tettensis.*
Flowering Season: Spring to early winter.
Native Range: Africa.
Comments: Parasol flower averages 6–8' tall and up to 10' wide, so this should give you a good idea about how much room to allocate for it in your garden. Pruning can keep it from dominating too much space, but keep your limb loppers sharpened, because this will need to be an annual task. It is not as aggressive as its close relative Chinese hat plant (*Holmskioldia sanguinea*), and it should not be confused with purple-flowered forms of that species. The flowers on *Holmskioldia* are curved and tubular (see Chinese hat plant in this guide), quite unlike the flowers of this species. Keep parasol flower on a regular fertilizer regimen, and give it chelated iron annually to avoid deficiencies.

Butterflies and hummingbirds find the blossoms of parasol flower irresistible, so if you have the room to dedicate to a large, sprawling shrub, then this should be a welcomed choice. Locating plants for sale will be much more difficult than Chinese hat plant, but you should be successful by checking specialty nurseries or plant sales held at botanical gardens, especially those sponsored by the Tropical Flowering Tree Society in Miami. Check the Internet for sources if all else fails.

Florida Shrub Thoroughwort

Koanophyllon villosum

Asteraceae (Aster Family)

Synonyms: *Eupatorium villosum.*

Flowering Season: Summer to fall.

Native Range: Southern Florida (Miami-Dade County) and the West Indies.

Comments: Another common name for this state-listed endangered species is Florida Keys thoroughwort, even though it does not occur naturally in the Florida Keys. It is a shrub to about 6' tall and can be locally common in pine rocklands and along the edges of hardwood hammocks in southern Miami-Dade County, including Long Pine Key in Everglades National Park.

It is a top-notch butterfly attractor but is grown in limited quantities by only a few local nurseries in southern Miami-Dade County that specialize in native plants. A well-grown specimen can be covered with thousands of off-white to pale violet flowers, to the delight of every butterfly within sight.

It will perform best in full sun but tolerates partial shade. It is puzzling that it has not garnered much horticultural attention because it fully deserves a Meritorious Achievement Award as a butterfly attractor. The related lateflowering thoroughwort (*Eupatorium serotinum*) has similar flowers and is also a superb butterfly attractor that is seldom cultivated.

Virginia Saltmarsh Mallow

Kosteletzkya pentacarpos
Malvaceae (Mallow Family)

Synonyms: *Kosteletzkya virginica.*

Flowering Season: All year.

Native Range: New York to Texas south through Florida, Bermuda, Greater Antilles, and southern Europe.

Comments: The singular beauty of Virginia saltmarsh mallow covered with pink blossoms is what inspires poets. It ranks high on my list of favorite wildflowers found in the vast marshes of the Everglades region. It is just as happy in freshwater wetlands as it is in saltmarshes. In garden settings it must have reliably wet soil, but regular irrigation will suffice if you do not have a personal wetland.

It reaches about 5' tall in prime conditions, and the 1½" flowers are produced off and on through all seasons of the year, peaking when summer rains flood its habitat. You will be hard-pressed to find it in cultivation, but nurseries that specialize in Florida native plants sometimes offer it. It can also be grown from seed collected outside of preserves or on private property with the landowner's permission.

Hummingbirds and butterflies visit the blossoms for nectar and pollen, so it is definitely a worthwhile addition to any garden.

Wild Sage

Lantana involucrata
Verbenaceae (Verbena Family)

Flowering Season: All year.

Native Range: Central and southern Florida, Bermuda, and tropical America.

Comments: This is the only Florida native lantana that should be widely cultivated, because it is common from Brevard and Hillsborough counties southward,
which covers the range of this guide. The yellow-centered flowers are white or
violet-tinged and are a favorite nectar source of butterflies and occasional hummingbirds. In the wild it lives in pinelands, scrub, and beach dunes, which are all
habitats that are not only dry but also subjected to periodic fires and salt spray, so
this is one tough plant. Also, it is not poisonous like the exotic *Lantana strigocamara*, and being a Florida native, it does not pose any environmental concerns.
Its only horticultural concern is that it may be prone to fungal leaf spotting with
overhead irrigation or if planted in an area of high humidity.

This is an upright shrub, but it often forms rounded mounds of stems to about
6' tall. It is hardy in tropical Florida and is widely available. Wild sage is attractive
in the landscape, especially when used in combination with other native plants
that share its habitat, such as saw palmetto (*Serenoa repens*), American beautyberry (*Callicarpa americana*), and coontie (*Zamia pumila*).

Lion's Ear

Leonotis leonurus
Lamiaceae (Mint Family)

Flowering Season: Summer into early winter.
Native Range: South Africa.
Comments: In its native Africa, the flowers of this colorful shrub are visited by sunbirds, but in Florida, the sunbirds are replaced by hummingbirds and orioles.

The softly hairy, orange flowers are produced in tiered whorls in the leaf axils and are quite showy in garden settings. The plant can reach 6' in height from seed in a single growing season, and in full sun or intermittent shade it will grow vigorously if given ample water. It is only moderately drought tolerant if that is a concern.

Hummingbirds are especially fond of the nectar produced by the flowers and will feed from them longer than they will at other blossoms. Large butterflies, especially swallowtails, also visit the flowers regularly.

Lion's ear is available from mail-order companies and from local specialty nurseries. It is winter hardy into Zone 8, but in colder climates it is grown as an annual. Use it as a border plant, or try it as a backdrop behind shorter plants. If kept in a container, it will require regular watering.

Christmas Berry

Lycium carolinianum
Solanaceae (Nightshade Family)

Flowering Season: All year.

Native Range: South Carolina to Texas, south through Florida and the West Indies.

Comments: Christmas berry is underused and underappreciated in tropical Florida gardens. In the wild, this tough plant grows along windswept, sandy and rocky seashores as well as in saltmarsh and mangrove habitats. Although it tolerates salty breezes and saline soil, it grows just fine in home gardens if given an exposed position in full sun. In winter it is often covered with decorative bright red fruits, so plant it near your patio or front door where you can augment its branches with ornaments around Christmastime.

Butterflies and hummingbirds visit the 4-lobed, lilac flowers that appear on long, wispy, often drooping branches. It is an uncommonly pretty shrub when flowering or fruiting, and its branching habit is quite unlike any other shrub you might grow.

You will find it available from nurseries that specialize in Florida native plants, or you can grow it from seed. Although they resemble tiny tomatoes, which are in the same family, the fruits of this species can cause vomiting if eaten. Still, they are not as toxic as many other members of the nightshade family.

Winged Loosestrife

Lythrum alatum var. *lanceolatum*
Lythraceae (Loosestrife Family)

Flowering Season: Summer.

Native Range: Virginia and North Carolina to Oklahoma and Texas, south through peninsular Florida.

Comments: This much-branched native shrub reaches 4–6' in height and is covered with an abundance of small violet flowers in summer. Its natural habitat is freshwater wetlands, so in a home garden it will require reliably wet soil or regular watering. Queens, viceroys, and monarchs are among the many butterflies that find the flowers irresistible; they sometimes swarm around flowering specimens as they wait their turn for a vacant flower.

Winged loosestrife has much horticultural potential, but it is seldom found in the Florida nursery trade. Propagation is by seeds, which can be collected from wild roadside plants. This should not be difficult, because it's common along ditches and canal banks. It is well worth growing, principally due to its bushy stature and its pleasing display of butterfly-attracting flowers. It can make an interesting subject as a container plant on a patio, or used as an addition to a water garden. This plant is the perfect example of a species that can be sought out, propagated, and included as something unique to your home garden. It is hardy throughout Florida.

Texas Waxmallow

Malvaviscus arboreus
var. *drummondii*
Malvaceae (Mallow Family)

Flowering Season: Spring into winter.

Native Range: Southeastern United States to Texas and Mexico. Naturalized in Florida.

Comments: Scottish naturalist Thomas Drummond first collected this winsome shrub in Texas around 1833, but decades passed before it was cultivated by gardeners in the Southeast. The Spanish name *manzanilla* (little apple) relates to the edible fruits, which taste like apples.

Texas waxmallow forms a compact shrub to about 3' tall but may reach more than twice that height in good growing conditions. Gardeners in cold temperate regions prefer to prune it back hard in wintertime to protect it from frost and to maintain it as a small shrub.

Relentless summer sun in tropical Florida will cause the leaves to be rough, small, and curled, so offer it the shade of a tall tree, and it will reward the hummingbirds and butterflies in your garden with a bounty of lovely red blossoms.

Seeds germinate promptly, and it roots easily from cuttings. A white-flowered form is available in nurseries, but it is not as attractive to hummingbirds. A related plant to consider is *Pavonia dasypetala;* it has very similar flowers.

Turk's Cap

Malvaviscus penduliflorus
Malvaceae (Mallow Family)

Flowering Season: All year.
Native Range: Mexico to Peru and Brazil.
Comments: Nodding red (or pink or white) flowers with closed petals adorn Turk's cap throughout the year. The flowers are a favorite of hummingbirds: rather than probing the length of the flower, they insert their beak through the base of the petals to access the nectar. This robs the flower of nectar by reaping that reward without performing the service of pollination. Many species of butterflies, especially hairstreaks and skippers, may be seen taking nectar that drips down the anther.

If you desire a compact shrub, Turk's cap requires dedication to pruning. It was once a popular landscape plant in Florida but is now mostly seen around older homesteads. It is trouble-free in the landscape and can either be used as a free-standing specimen or given a fence to ramble upon. The stems root wherever they touch the ground; this should be discouraged, or you will inherit a long weekend chore to gain any semblance of control. It is not as easy to find in nurseries as it once was, but it roots very easily from cuttings if you know someone who grows it. A mature specimen is uncommonly ornamental.

Snow Squarestem

Melanthera nivea

Asteraceae (Aster Family)

Flowering Season: All year.

Native Range: Illinois and Kentucky south along the Gulf and Atlantic Coastal Plain throughout Florida.

Comments: There are two very disparate growth forms of this species in Florida, and some botanists have attempted to separate them as different varieties or even distinguish them as two distinct species. The form found in the Everglades region has a low, sprawling growth habit with long trailing stems and is sometimes treated as an endemic species called *Melanthera parvifolia*. To the north of this region, it takes on a bushy, shrublike habit; it is this form that looks best in gardens, but you should grow whatever form is available in your region.

The flowers have black anthers and are in rounded clusters at the branch tips. The flowers are visited by an array of butterflies, especially hairstreaks, crescents, blues, and skippers. The bushy form prefers full sun and sandy, well-drained soils. The sprawling form from the Everglades is often found in wet prairies growing in the gray, alkaline, claylike soil called marl.

Check nurseries that specialize in Florida native plants, or search for sources on the Internet. Both forms of snow squarestem are perfectly adaptable to container culture and are lovely when in flower.

Wooly Teabush

Melochia tomentosa
Malvaceae (Mallow Family)

Flowering Season: All year.

Native Range: Florida (Miami-Dade and St. Lucie counties), Texas, West Indies, and tropical America.

Comments: The name wooly teabush comes from the West Indies, where its fuzzy leaves are used as a soothing tea. A well-grown specimen is quite charming in garden settings when it is covered with its small, hibiscus-like, pinkish-violet flowers. Butterflies, hummingbirds, and bees are very fond of the blossom nectar and will be the center of attention throughout the year.

Until recently, the genus *Melochia* was placed in the cacao family (Sterculiaceae), which also included *Theobroma cacao,* the source of chocolate. The former cacao family is now included in the mallow family. Wooly teabush prefers full sun and alkaline soils that drain readily. It may reach 6' tall in proper growing conditions.

Finding plants for sale will take some diligence at present, but check plant sales at Fairchild Tropical Botanic Garden in Coral Gables. Hopefully, once the word gets out, growers that specialize in Florida native plants will begin propagating it, for it fully deserves horticultural attention for Florida gardeners. Propagation is by seed or cuttings. Another native species to consider is the more common bretonica peluda (*Melochia spicata*).

Crimson Beebalm

Monarda didyma
Lamiaceae (Mint Family)

Flowering Season: Late spring through summer.
Native Range: Eastern United States south to Georgia.
Comments: Crimson beebalm is well-known to gardeners, especially within its native range. It is also commonly cultivated in the northern half of Florida, but it is seldom seen in the southernmost counties even though it grows quite well throughout Zone 10. Because it is native as far north as Maine, it goes without saying that it does not need winter protection in Florida.

The flowers look as if they are shouting to announce their delectable nectar, and every hummingbird in your garden will attempt to defend the territory around a flowering specimen. The flowers also attract butterflies and are favorites of native bees, hence the name beebalm. It prefers full sun and soil that retains moisture, either in the ground, a raised bed, or in a container where it receives adequate air circulation to avoid powdery mildew. It will mature between 3–4' tall; in garden settings, it is best used among other plants of similar stature.

For color in your garden, you will not be disappointed with crimson beebalm. It is offered by nurseries from Central Florida northward and is widely available from Internet sources. Two popular cultivars are 'Pardon My Pink' and 'Jacob Cline.'

Spotted Beebalm

Monarda punctata

Lamiaceae (Mint Family)

Flowering Season: Late spring through summer.

Native Range: Eastern United States south into Florida.

Comments: Unlike the previous species, spotted beebalm prefers dry soils and has purple-spotted white or yellow flowers with pink bracts produced in whorls at the leaf axils. It will prefer a sunny spot with good air circulation and where hummingbirds and butterflies have unrestricted access to the flowers.

In good growing conditions, it may reach 5' tall, but its average height is around 3' with an equal spread. Its natural range extends into lower Central Florida, but it can be successfully grown throughout South Florida wherever there is sandy, well-drained soil. It is small enough to be grown in a container on a sunny patio or in a raised bed filled with a good cactus soil mix that drains quickly.

The insect activity around flowering specimens also attracts the attention of insect-eating birds. If you should get thirsty while pulling weeds in your garden, you will be pleased to know that the leaves make a refreshing tea and have been used medicinally by Native Americans for centuries. Plants and seeds are available through Internet sources if you cannot find a local source.

Ornate Banana

Musa ornata

Musaceae (Banana Family)

Flowering Season: Early spring through late fall.

Native Range: Southeast Asia.

Comments: If you would like to add a distinctive tropical ambience to your garden, then this ornamental banana will be happy to please. The male flowers attract hummingbirds and other nectar-seeking birds such as Baltimore orioles and the Central American spot-breasted oriole that is naturalized in the Miami area.

Ornate banana prefers rich soil and benefits from an annual topping of composted cow manure or kitchen compost. It reaches just over 6' tall, and unlike edible bananas, the flowering and fruiting spikes are held vertically. The inedible violet fruits are decorative and produce very hard seeds that can be planted or given to fellow gardeners. After female flowers have been pollinated, the plant continues to produce male flowers over a period of months. After a row of them is spent, the fleshy bract falls off and another one opens to reveal fresh flowers.

Like all bananas, once a stem has finished fruiting, it dies, but during that process it will produce offsets at the base. Clumps should be thinned occasionally to avoid overcrowding. The flowers of edible bananas also attract hummingbirds if you have the space.

Queen Sirikit

Mussaenda erythrophylla ×
Mussaenda philippica
Rubiaceae (Madder Family)

Flowering Season: Periodically all year.
Native Range: Of hybrid origin.
Comments: The large, garish bracts of this hybrid outshine the small, star-shaped, orange to yellowish-orange flowers; they are a ploy by the plant to catch the attention of potential pollinators and direct them to the true blossoms. One parent of this plant has brilliant red bracts with orange flowers, and the other has pure white bracts with orange flowers. The hybrid was named to honor Queen Sirikit of Thailand for her visit to the Philippines in the 1970s.

This shrub reaches 8' tall or more; it tends to get rangy looking as it matures, so seasonal pruning is in order to keep it compact. It is not drought tolerant, so plant it where it can be watered when necessary. It will tolerate high, shifting shade but prefers full sun through much of the day.

Hummingbirds and butterflies visit the flowers. In the landscape, mussaendas are like loud children in a library—with their vibrant colors, they almost scream for attention—so plant them where their bold appearance will be welcomed and not too overwhelming.

Moujean Tea

Nashia inaguensis
Verbenaceae (Verbena Family)

Flowering Season: All year.

Native Range: Little Inagua and Great Inagua, Bahamas.

Comments: When crushed, the tiny leaves of this Bahamian shrub are strongly aromatic with an odor reminiscent of dish soap. The flowers, though small, are very appealing to butterflies and small, day-flying moths.

In the landscape, moujean tea must have full sun. It will tolerate drought conditions once it becomes established, but it is extremely intolerant of the soil drying out in a pot and will quickly perish if it does not receive regular watering. Thus it is not suitable for container culture, so plant it in the ground soon after you arrive home with your prized possession. In the garden, it will reach about 6' tall with crisscrossing branches.

Because it is found naturally only on two small Bahamian cays, it has only recently been introduced into cultivation—for this, we can thank Fairchild Tropical Botanic Garden in Coral Gables. Moujean tea is quite salt tolerant and will withstand sea breezes in coastal situations. It is hardy into the warmer regions of Zone 9 and is available at Fairchild Tropical Botanic Garden plant sales or at specialty nurseries in the southeasternmost counties. If they are not convenient, then check Internet sources.

Shellflower

Ocimum labiatum
Lamiaceae (Mint Family)

Synonyms: *Orthosiphon labiatus.*
Flowering Season: Spring into early winter.
Native Range: Africa.
Comments: Shellflower is a very popular landscape plant in South Africa, where it excels at attracting butterflies; but Africa doesn't have hummingbirds, and we do, and they can hardly maintain their composure around a flowering specimen.

In the absence of pruning, expect shellflower to reach 6' tall with an equal spread, forming a rounded mound of stems blanketed with lilac flowers. It has no special garden needs other than full sun to light shade, with occasional pruning to help maintain a tidy shape. Although it is drought tolerant, it looks much nicer when it receives regular watering in the dry season.

Shellflower is gaining popularity in Florida, but you may have to rely on specialty nurseries, botanical garden plant sales, or Internet sources to find plants for sale. It is a genuine treat for butterflies and hummingbirds, so put it at the top of your gardening wish list. It is known to be hardy down to 30°F, so little winter protection in tropical Florida should be necessary. It is wise, nonetheless, to cover it with a blanket if temperatures close to freezing are forecast.

Purple Firespike
Odontonema callistachyum
Acanthaceae (Acanthus Family)

Flowering Season: Fall through winter.
Native Range: Mexico and Central America.
Comments: When purple firespike is in full regalia, it becomes the center of attention in any garden with its tall, terminal spikes of tubular, pinkish-lilac flowers that beckon to hummingbirds and butterflies alike. It can be successfully grown in full sun or transitional shade and forms a rounded mound of stems to about 6' tall if left on its own. It is widely available from nurseries throughout tropical Florida.

Once purple firespike is established, it will require little care other than removing stems that grow out of bounds. It can be pruned during the growing season to encourage new growth and to reduce its overall size if necessary. Give the cut stems to fellow gardeners, because they root very easily.

Light applications of fertilizer with micronutrients in late spring will help avoid any nutrient deficiencies during the rainy season. Place it in the garden as a backdrop for shorter plants or let it claim a corner of your property. It looks particularly attractive when planted next to the following species.

Red Firespike

Odontonema cuspidatum
Acanthaceae (Acanthus Family)

Synonyms: *Odontonema strictum.*
Flowering Season: All year.
Native Range: Central America.
Comments: Red firespike is one of those remarkable plants that displays continuous year-round color coupled with ease of care. It excels in full sun or filtered light and is an unconditional must for your garden. The glistening crimson blossoms are produced in terminal, often branching spikes and are so shiny they appear to be waxed. It typically reaches 6' tall with many erect stems, sometimes arching to the ground, where they take root. The plant prefers rich soil; when grown in full sun, it will benefit from a layer of mulch to help retain soil moisture.

Red firespike is one of the favorite flowers of ruby-throated hummingbirds, which will zealously guard the territory around the plant from intruders. Butterflies of all creeds and colors are continuous sights around a flowering specimen as well.

It is widely available and can be used in groups to create bright color beneath tall, open-canopied trees. Propagation is from cuttings or by dividing established plants. A similar species grown by collectors is coral firespike (*Odontonema tubaeforme*), which has larger, coral-colored flowers. 🐦 🦋

Cat Whiskers

Orthosiphon aristatus
Lamiaceae (Mint Family)

Synonyms: *Orthosiphon stamineus.*
Flowering Season: All year.
Native Range: China.
Comments: Cat whiskers typically produces erect, terminal spikes of violet flowers, but a white-flowered form is also commonly cultivated. When grown together they make interesting ornamental garden subjects. The long, upward-curving stamens do, indeed, resemble the whiskers of a cat. This is a low shrub that only reaches 3–4' tall at maturity. It can be used as a border plant in full sun or in mass plantings of mixed flower colors to achieve a tall groundcover. It can also be mixed with other plants of equal height, such as pentas (*Pentas lanceolata*) or tropical sage (*Salvia coccinea*). Giant swallowtails are especially fond of the flowers, and sphinx moths visit the blossoms as well.

Cat whiskers needs full sun and may require supplemental drenches of chelated iron in the heavy alkaline soils of the southernmost Florida counties. Otherwise it is rather trouble-free in the landscape and is a good candidate for a container to decorate a sunny porch or patio. Propagation is by seeds or cuttings, and it is often available in department store garden centers. It may survive mild winters in warmer sections of Zone 9.

Golden Shrimp Plant

Pachystachys lutea
Acanthaceae (Acanthus Family)

Flowering Season: All year.
Native Range: Peru.
Comments: The brilliant yellow bracts and snow-white flowers of this 3–4′ plant create a fabulous effect in gardens, offering a flashy floral display throughout the year. Golden shrimp plant is extremely versatile, whether in spectacular mass plantings, as a colorful border along walkways, or as an ornate container plant for a sunny porch or patio. Mix it with other plants of equal stature to create a colorful and enticing area for both hummingbirds and butterflies. It provides color throughout all seasons of the year, so give it a prominent position in the garden. Mass plantings of golden shrimp plant are nothing short of spectacular.

In the southern counties of Florida, where hummingbirds are only present into May, it is helpful to cut stems back in late spring to create bushy plants loaded with flowers the following fall, just as migration begins anew. Sunny locations are best, but it performs well where it only receives afternoon sun or an eastern exposure with morning and midday light. It is widely available from nurseries and is so popular it can even be found in department store garden centers.

Feay's Palafox

Palafoxia feayi

Asteraceae (Aster Family)

Flowering Season: Late summer into spring.

Native Range: Florida.

Comments: Feay's palafox has a natural range that extends from Volusia and Marion counties south to Collier and Miami-Dade counties. It is absent from Everglades National Park because it prefers sandhill and scrub habitats not found in the southern Everglades region.

This leggy species may stand about 5' tall, with clusters of attractive flowers on long stems arising from the leaf axils and branch tips. A flowering plant in its native habitat will have a riot of activity around it, ranging from nectar-seeking bees, wasps, butterflies, and diurnal moths, which in turn attract warblers, vireos, gnatcatchers, and flycatchers looking to cash in on the bounty. Green lynx spiders often hide beneath the flowers waiting to grasp a hungry insect that shows up for a nectar feast.

Look for this species in nurseries that specialize in Florida native plants, or check Internet sources. Full sun and sandy, well-drained soil are what it requires from gardeners. It is very drought tolerant and generally does not require any supplemental irrigation once it is established. If you are able to find this species, consider it a prized possession.

Pentas

Pentas lanceolata
Rubiaceae (Madder Family)

Flowering Season: All year.

Native Range: Eastern Africa to southern Arabia.

Comments: It is tempting to recommend planning your butterfly garden around this species, not only because it is available in an array of radiant colors but also because butterflies and hummingbirds visit the flowers and come back for second helpings. Pentas, also called starflower, is one of the most popular plants in Florida, and with good reason. Flowers come in shades of red, pink, white, and violet, and mass plantings of different colors are nothing short of spectacular. Some forms reach about 4' tall, while others mature at half that height.

Pentas should not be watered unless it is wilting, and automatic overhead irrigation should be avoided because it can cause leaf fungus. This plant thrives in dry soil in full sun but will tolerate transitional shade.

It also excels as a container plant for sunny areas, and the low-growing forms can be used in window boxes or planters. If you want your butterflies and hummingbirds to be happy, then this is a "must-have" for tropical Florida gardeners. You may also find larvae of the tersa sphinx moth eating the leaves. They're really cool moths, so leave the larvae alone.

Camelia Rosa
Pereskia portulacifolia
Cactaceae (Cactus Family)

Flowering Season: Spring through fall.

Native Range: Hispaniola.

Comments: Looks can be deceiving: do not be outwitted by the lush leaves and stunning pink flowers, because this shrub has wickedly sharp spines just like most every other cactus. Camelia rosa, or pink rose cactus, grows in dry forests and coastal scrub in its native Haiti and Dominican Republic. It is imperiled due to habitat loss and currently occurs in only four known wild populations.

In cultivation, it requires full sun and dry, gravelly soil and should be planted where people and pets will not be exposed to the needle-sharp spines. It can reach tree stature in the wild, but in Florida it is typically shrubby with long branches intertwined to create a pretty, but formidable, specimen. Hummingbirds pollinate the flowers as they seek both nectar and high-energy pollen.

Camelia rosa can be kept in a container, if you prefer not to grow it in the ground; just fill a decorative pot with any standard cactus soil mix and go light on the water. It is available from Internet sources and from specialty nurseries. Check the South Florida Cactus and Succulent Society's plant sales for this and other *Pereskia* species. 🐦

Sweetscent

Pluchea odorata
Asteraceae (Aster Family)

Flowering Season: Summer and fall.

Native Range: Eastern United States, Bermuda, and tropical America.

Comments: This is one of the best examples of a Florida native plant that attracts multitudes of butterflies with its charming clusters of pink flowers yet has never received horticultural attention. It is fast growing, and seeds should be collected and planted every few years because the plant is somewhat short-lived. It requires full sun and will tolerate both wet and moderately dry conditions, plus it is salt tolerant.

Sweetscent is commonly found in saltmarsh habitats and freshwater wetlands but grows just fine in home garden settings. Seeds can be gathered in fall and winter in unprotected areas. The leaves of this species are very aromatic when crushed, and the flowers are fragrant. Its typical height is about 3–4', so it is useful as a border plant, or it can be tucked in among other plants of similar or lower stature. It would also be an interesting subject for a container on a sunny porch or patio. Other choices are the large, bushy cure-for-all (*Pluchea caroliniensis*) and the small, herbaceous rosy fleabane (*Pluchea baccharis*). All members of the genus have been used medicinally and placed in bedding to deter fleas.

Blue Plumbago

Plumbago auriculata
Plumbaginaceae (Plumbago Family)

Synonyms: *Plumbago capensis.*
Flowering Season: All year.
Native Range: South Africa.
Comments: Alluring blue flowers cover this shrub throughout the year. The flowers perfectly match the dorsal wing color of the small Cassius blue butterfly, which not only visits the flowers for nectar but also uses it as a larval host plant. The feeding activities of the minuscule larvae will never be noticed, so this is of no concern regarding aesthetics in the landscape. It's difficult enough just to find one of the larvae.

Pictured is a cultivar introduced in the 1990s with much bolder blue flowers that stand out in the landscape. A white-flowered form is also cultivated, but do not mistake it for the white-flowered native *Plumbago zeylanica*. Blue plumbago is commonly seen in mass border plantings and is popular in Florida gardening circles. It grows to a height of about 5' and prefers to be kept dry. Overhead irrigation may cause leaf fungus problems, so this should be avoided.

Blue plumbago is exceptionally attractive, and mixing the blue-flowered and white-flowered forms creates a colorful design. Let the red-flowered *Plumbago indica* scramble around among them if you are feeling particularly patriotic.

Proctor's Bellflower

Portlandia proctorii
Rubiaceae (Madder Family)

Flowering Season: Sporadically all year.

Native Range: Jamaica.

Comments: This quaint and charming shrub produces glamorous red tubular flowers that measure just over 2" long and about ¾" wide. In Florida, it typically attains a height of 5–6' with a shrublike growth habit.

Your best chance of acquiring this species is to attend plant sales held at Fairchild Tropical Botanic Garden, especially the sale sponsored by the Tropical Flowering Tree Society. Check the events calendar on the garden's Web site. It is only hardy on the mainland in Zone 10B, and even there it will require winter cold protection if temperatures are expected to reach 50°F or below. If it's in a pot, bring it indoors.

It is an acclaimed hummingbird attractor in Jamaica, and the ruby-throated hummingbird delights in its flowers here in Florida. Not to be outdone, skipper butterflies are adept at crawling down the floral tube to reach the nectar.

The species name honors botanist George Proctor, who has worked on the flora of Jamaica for more than fifty years. He was imprisoned in Jamaica in 2010 for conspiracy to murder his wife and three other women and was released in 2012 at the age of ninety-one.

White Wings

Pseudomussaenda flava
Rubiaceae (Madder Family)

Synonyms: *Mussaenda flava.*
Flowering Season: All year.
Native Range: Equatorial Africa.
Comments: A snow-white leaflike floral bract embellishes the faintly perfumed, starlike, yellow flowers of this elegant shrub. This is one of several plants that produce large, bright floral bracts to help attract the attention of pollinators to their small flowers. White wings may reach 4–5' tall and is more tolerant of dry soils and cool winter temperatures than its close relatives in the genus *Mussaenda,* and it is cold hardy into Zone 9.

White wings prefers full sun or very light shade and can be used as a freestanding specimen or planted on the sunny side of other shrubs. It blooms throughout much of the year and is attractive to both butterflies and hummingbirds.

White wings will not likely be found in mainstream garden centers but is available from nurseries that specialize in flowering plants, or through Internet sources. It can be kept in a decorative container for a deck or patio, but it must have soil that drains well. Perlite or coarse gravel will aid in soil drainage. Check botanical gardens if you would like to see mature specimens in a garden setting.

Wild Coffee

Psychotria nervosa
Rubiaceae (Madder Family)

Flowering Season: Spring and summer.

Native Range: Florida and throughout much of tropical America.

Comments: This species is the most common and widespread member of its genus in Florida. The clusters of small white flowers are attractive to butterflies, especially the ruddy daggerwing. It is very versatile in the landscape and is best used in small groups planted in partial shade beneath tall trees. Because its ultimate height is about 6', it is useful as a privacy screen. It is an understory shrub in tropical hammocks, so it performs best in transitional shade or where it only receives full morning sun. The glossy, dark green leaves are complemented by bright red, fleshy fruits that are a favorite food of northern mockingbirds and gray catbirds.

One potential horticultural issue is the larvae of a leaf-roller moth, which fold the leaves into a tube, forming shelters in which they hide and pupate. This is seasonal, and no control is recommended because pesticides used to kill the larvae will also kill butterflies. Just let nature run its course.

Two related native species, *Psychotria sulzneri* and *Psychotria ligustrifolia*, can also be used but are less commonly cultivated. The genus name, *Psychotria,* is Greek for "spirit" and alludes to Neotropical species with psychoactive properties.

White Indigoberry
Randia aculeata
Rubiaceae (Madder Family)

Flowering Season: All year.

Native Range: Central and southern Florida and tropical America.

Comments: The fragrant flowers of white indigoberry are a choice nectar source for the rare and beautiful Atala butterfly of southeastern Florida. Although it can attain small-tree stature, to 12' or more, it is usually shrubby, from 4–6' tall, with a small, tidy canopy. Its natural habitat is sunny pinelands, but it is also a midstory tree in hardwood forests, where it grows in dappled sunlight. It is fairly well-known among native-plant enthusiasts in southern Florida and is often used in plantings with other native trees and shrubs to simulate a native habitat. The pulp inside the white fruits is purple and has been used as a dye.

White indigoberry is available from nurseries that specialize in native plants and is hardy in Zones 10 and 11. Although some plants produce sharp thorns, it is usually thornless, with a somewhat angular branching habit.

It is small enough to be maintained for some time in a container, but be sure the container is large enough to keep the plant from becoming root-bound.

Firecracker Plant

Russelia equisetiformis
Plantaginaceae (Plantain Family)

Flowering Season: All year.
Native Range: Mexico.
Comments: The cascading stems and coral-red flowers of this popular land-scape plant create a fountain of color. Butterflies and hummingbirds are common around a flowering plant. Forms with pink, creamy yellow, or white flowers are available.

Although it has become rather passé due to the introduction of newer species, it still looks exceptionally bold when planted where the stems can cascade over rocks or down a wall. Over time, the plant may become congested with stems, so be prepared to prune it back severely, even to the point of cutting the stems off at ground level. It will rebound quickly with fresh new stems.

Firecracker plant forms mounds to about 4' tall and blooms freely until winter cold triggers dormancy. In warm years, it may flower throughout winter in the warmer regions of Zone 10. It can be kept in a container, provided it has full sun and dry soil. It is quite spectacular when planted in masses and can even be used as a low hedge. Planting the various color forms together is quite striking and will be sure to draw comments from houseguests.

Desert Fire

Russelia rotundifolia
Plantaginaceae (Plantain Family)

Flowering Season: All year.

Native Range: Mexico.

Comments: There is a cultivar of this species called 'San Carlos' that is also sometimes listed as a cultivar of *Russelia coccinea*. Regardless of which name is correct, this species is one of the ultimate hummingbird attractors, and butterflies find the flowers to be irresistible as well.

Once it is established, desert fire thrives in heat and drought, so this rugged plant will work in areas where other plants might fail. It is an outstanding border plant for lining entranceways, delineating the edges of formal gardens, or forming a colorful border along a sunny patio. It can even make an interesting subject for a hanging basket. It is very versatile, so try it in a combination of situations until you find what works best.

Stems reach 3–4' long and arch gracefully to form a mound. If plants have personalities, then desert fire is the life of the party. It is becoming more popular, but if you cannot find it locally, it is widely available on the Internet. Winter hardiness extends to Zone 8B, so there is no need for cold protection in Zones 10 or 11.

Red Rocket

Russelia sarmentosa

Plantaginaceae

(Plantain Family)

Flowering Season: All year.

Native Range: Mexico and Central America.

Comments: Your third choice in the genus *Russelia* is red rocket, but the hummingbirds and butterflies will not be disappointed if you plant all three. This sun-loving plant produces multiple, long, arching stems with rounded, axillary whorls of small but brilliant red flowers evenly spaced along the stems.

It is versatile in the landscape and can be scattered here and there in gaps between other flowering plants. It can also be planted in beds around trees. If you have limited space, then try it in a hanging basket, allowing its long, arching branches with cheery red flowers to cascade. In addition to its flowers, which appeal to hummingbirds and butterflies, its leaves serve as larval food for the common buckeye butterfly.

Red rocket is relatively easy to find in garden centers and nurseries, plus it is widely available on the Internet. It is prone to nutrient deficiencies in limestone soils, so a light application of fertilizer is recommended at the beginning of the rainy season and in early fall.

Azure Blue Sage
Salvia azurea
Lamiaceae (Mint Family)

Flowering Season: Summer into winter.
Native Range: Eastern and midwestern United States.
Comments: There are so many *Salvia* species from which to choose that it was difficult to decide which to include in this guide. Yet azure blue sage is a must, because it is clearly a winner for both hummingbirds and butterflies. It may reach 5' tall with spikes of whorled, azure, ¾" flowers. To maintain its shape, prune it moderately in early springtime. This species ranges through the southeastern United States into Central Florida and often produces flowers with 2 parallel white lines on the lip.

It is hardy northward into Zone 5, so it is decidedly cold tolerant. Keep this plant on the dry side, where it receives sun all day, or the stems will become lanky and topple easily. It typically begins blooming when summer is in full swing and continues into winter in tropical Florida. Although hummingbirds and butterflies are frequent visitors, bees pollinate the flowers in the wild. Some taxonomists refer to a large-flowered form as var. *grandiflora.*

Wendy's Wish

Salvia 'Wendy's Wish'
Lamiaceae (Mint Family)

Flowering Season: All year.

Native Range: Of cultivated origin.

Comments: The best reason to plant this sensational sage is that proceeds from sales go to the Make-A-Wish Foundation, which grants the wishes of children with life-threatening medical conditions. Australian gardener Wendy Smith first discovered this plant in her garden, and although it is unclear if it is a cultivar or a hybrid, it is generally believed that a Mexican species, *Salvia buchananii,* figures into its origin, because the original seedling appeared among plantings of that species in Wendy's garden. It was introduced into the United States in 2009 and patented in April 2011. Its popularity soared so rapidly that you can now find it in department store garden centers.

Hummingbirds and butterflies go into a tizzy when this plant shows off its striking blossoms, which is often. It reaches 3' tall, preferring full sun to very light shade, and makes a glamorous border plant around a patio or lining a walkway.

Here is your chance to not only make the hummingbirds and butterflies in your garden happy but also help make a child's wish come true. 🐦 🦋

Tropical Sage

Salvia coccinea

Lamiaceae (Mint Family)

Flowering Season: All year.

Native Range: Southeastern and south-central United States, Bahamas, West Indies, and Mexico south into South America.

Comments: Hardly any introduction is necessary for tropical sage, because it is hugely popular among gardeners for its sparkling red flowers that are present throughout the year. White- and pink-flowered forms are also commonplace, and there are quite a number of named cultivars available in the trade. Hummingbirds and butterflies frequent the flowers, and painted buntings, indigo buntings, and several kinds of finches feast on the seeds.

It reaches 3–5' tall and spreads freely from seed wherever there is bare soil, but most gardeners simply allow it to move about wherever it prefers. It is often used in mass plantings but can be incorporated among other flowering plants. It prefers full sun for maximum flowering but will tolerate high, shifting shade. It is available from many Florida nurseries and also from Internet growers, which offer both plants and seeds. This is another one of those "must haves" for anyone interested in attracting hummingbirds and butterflies with native plants.

Blue Anise Sage

Salvia guaranitica
Lamiaceae (Mint Family)

Synonyms: *Salvia coerulea.*
Flowering Season: Spring into early winter.
Native Range: South America.
Comments: Hummingbirds and butterflies head straight to the flowers of this semiwoody species with anise-scented leaves. It has a rounded growth habit to 5' tall or more and is covered with tall spikes of blue flowers throughout much of the year. There are numerous named cultivars, including 'Black-and-Blue,' with a nearly black calyx (pictured), and 'Argentine Skies,' with attractive pale blue flowers.

This sage succeeds best in full sun but will tolerate partial shade. In gardens it looks especially appealing when mixed with other members of the genus, so try it in combination with the red-flowered tropical sage (*Salvia coccinea*), 'Hot Lips' sage (*Salvia microphylla*), or even golden shrimp plant (*Pachystachys lutea*). Wherever you plant it, you will be pleased by its sensational floral display, with the added bonus of butterflies and hummingbirds darting among the flowers.

Pruning the plant in late spring will encourage new growth and a more compact form. It is quite cold tolerant, surviving without damage winter temperatures of 40°F in our Homestead, Florida, garden.

Forsythia Sage
Salvia madrensis
Lamiaceae (Mint Family)

Flowering Season: Sporadically all year.

Native Range: Mexico.

Comments: Members of this genus are excellent at providing splashes of color in the garden, and forsythia sage does not disappoint. The stout, square stems are lined with fragrant, textured leaves, and each stem is topped with a tall spike of tubular yellow flowers. The plant typically reaches 4–6' tall, and older stems tend to lean outward while new stems emerge from the center of the plant. Stems root wherever they touch ground, so it will require periodic thinning to keep it in bounds.

Hummingbirds are exceptionally fond of the flowers, and so are a parade of butterflies. Because it grows at high elevations in the tropics, it can tolerate winter temperatures even where it freezes. A cultivar named 'Dunham' has survived -9°F in North Carolina.

Forsythia sage succeeds in full to partial sun, but give it plenty of room. Due to its height and long-lasting displays of showy flowers, it lends itself well as a backdrop for shorter plants. A cultivar called 'Red Neck Girl' has burgundy stems and is used in the cut flower trade. Check nurseries that specialize in butterfly-attracting flowering plants. It is also readily available from Internet sources.

'Hot Lips' Sage

Salvia microphylla cv. 'Hot Lips'
Lamiaceae (Mint Family)

Flowering Season: All year, peaking in spring and fall.

Native Range: Mexico and southeastern Arizona.

Comments: Typical *Salvia microphylla* has red flowers, but the cultivar 'Hot Lips' produces solid red, pure white, and two-toned red-and-white flowers on the same plant. It reaches about 4' tall under good growing conditions and is so free-blooming it is a sheer delight to have in any hummingbird and butterfly garden. The two-toned flowers are absolutely darling and will make for a nice conversation piece when you have houseguests.

Give this species full sun with well-drained soil, and it will thrive with little care. Seedlings may or may not produce two-toned flowers. The leaves smell strongly of mint and blackcurrants; in Mexico they are used for brewing herbal tea. There are more than 20 named color forms and hybrids of this species in the nursery trade, so check Internet sources to find ones that meet your fancy. 'Hot Lips' is sometimes found in local nurseries in Florida. Some sources (incorrectly) call it *Salvia × jamensis*. 🐦 🦋

Belize Sage
Salvia miniata
Lamiaceae (Mint Family)

Flowering Season: Spring into early winter.
Native Range: Mexico and Belize.
Comments: When I first saw this species in flower, I knew it would be a butter-fly and hummingbird magnet, and it wasn't long before it proved my hunch. The fuzzy, intensely red flowers are exceptionally pretty, and a single plant can form a mound of stems to 4' tall with an equal width. Stems may root where they touch ground, so you might grow a large patch of it before long. Dig up clumps of it and give them to your gardener friends. It can tolerate high, shifting shade, so here is your chance to add color to your not-so-sunny areas. It is grown as an annual in cold temperate regions but will need no winter protection in tropical Florida.

Belize sage prefers rich soil that has been amended with compost or compos-ted cow manure, and during dry spells it benefits from deep, weekly watering. A blooming patch of this species is inordinately beautiful, so put it on your list of plants to acquire. Check botanical garden plant sales and Internet growers that specialize in this genus.

Other noteworthy red-flowered species are the Bolivian *Salvia oxyphora* and the Mexican *Salvia involucrata*.

'Van Houttei Peach' Sage

Salvia splendens
'Van Houttei Peach'
Lamiaceae (Mint Family)

Flowering Season: Summer into winter.
Native Range: Brazil.
Comments: *Salvia splendens* has produced some of the most commonly grown *Salvia* cultivars, with color forms sporting yellow calyces with white flowers, orange calyces with pink flowers, magenta calyces with magenta flowers, and peach calyces with peach flowers (pictured), to name a few. The typical wild species produces brilliant red calyces with matching red flowers. It reportedly grows from 5' to more than 20' tall, but Dutch growers made dwarf varieties that only reach 15" in height. The cultivar 'Van Houttei Peach' is a tender perennial that reaches about 20" tall. It makes an outstanding bedding plant, and even though the peachy floral colors are quite subtle, it will be the talk of your garden when it flowers and every hummingbird within flying distance stops by to enjoy the nectar. Large butterflies like swallowtails, sulphurs, and fritillaries will be there also.

Give this plant full sun to light shifting shade, either in the ground or in a container. It looks perfectly stunning when planted in beds by itself or mixed with other colors.

Sanchezia

Sanchezia speciosa
Acanthaceae (Acanthus Family)

Flowering Season: All year.

Native Range: Possibly Ecuador or Peru; known only from cultivation.

Comments: Sanchezia offers a distinct tropical flair with its large, dark green leaves adorned with white or bright yellow veins. The plant produces multiple stems that reach 6–8' tall and are topped with spikes of reddish-orange bracts and tubular yellow flowers. Add a hummingbird zipping from flower to flower, and the picture is complete.

Sanchezia spreads by rhizomes and will develop into a colorful cluster of stems over time. The stems are sparingly branched near the top, but pruning will encourage further branching if desired. The leaves may reach nearly 2' long in shaded areas, but the plant looks best if it receives full sun during the morning, with shifting shade during the heat of midday. It performs perfectly well next to a wall or when used as an accent plant beneath tall trees. Pinnate-leaved palms offer the ideal amount of shade.

It is hardy in Zones 10 and 11, but further north it may get killed to the ground in winter, resprouting from the base in spring. It prefers moist, humus-rich soil and requires regular watering until it becomes firmly established. *Sanchezia nobilis* is similar but has smaller leaves and wider flowers.

Golden Plume

Schaueria calicotricha

Acanthaceae (Acanthus Family)

Synonyms: *Schaueria flavicoma.*

Flowering Season: Mostly late summer to early winter in Florida.

Native Range: Brazil.

Comments: The curved, tubular flowers of this 3–4' perennial shrub are lemon yellow and surrounded by interesting clusters of feathery, pale yellow to yellowish-green bracts. It is easy to grow and performs as well in dappled light as it does in full sun. Once it begins flowering, it continues for many weeks, much to the delight of ruby-throated hummingbirds and an array of nectar-seeking butterflies.

Although it may be difficult to find local sources in Florida, golden plume is widely available on the Internet from a host of mail-order sources.

It is perfectly adaptable to container culture, where it favors a rich, well-drained soil mix. With judicial pruning in spring, it will form a dense, rounded shrub with clusters of flowers at every branch tip, and it will be the center of attention in your garden. Keep your clippers handy, because your gardener friends are going to want cuttings.

In cooler zones it may get killed to the ground from frost or freezes, but it should respond quickly when warmer weather returns in spring.

Blue Lips

Sclerochiton harveyanus
Acanthaceae (Acanthus Family)

Flowering Season: Summer to winter.

Native Range: South Africa's Eastern Cape north to Zimbabwe.

Comments: The common name of this engaging plant is taken from *blou lippe* in Afrikaans, a Dutch-derived language widely spoken in South Africa. In its native habitat, blue lips can become a small tree to 12′ tall, but in Florida it typically takes on a shrubby appearance if planted on its own. It prospers when planted near a tree, where it will climb into the lower canopy and then poke its stems out into the sun to flower.

Unusual and alluring blue flowers begin their show in summer and continue well into winter. Take note, gardeners, because the flowers are attractive to both butterflies and hummingbirds. Blue lips lives in African forests, so plant it where it receives dappled sunlight in rich, well-drained soil. Nurseryman Jesse Durko of Broward County introduced blue lips from Mozambique during one of his African excursions.

Specialty nurseries in Miami-Dade and Broward counties grow it, but if you have difficulty finding it locally, then check Internet sources. The flowers will be a constant source of conversation.

Bahama Senna

Senna mexicana var. *chapmanii*
Fabaceae (Pea Family)

Synonyms: *Cassia bahamensis,* misapplied; *Cassia chapmanii.*
Flowering Season: All year.
Native Range: Southern Florida, Cuba, and Bahamas.
Comments: Bahama senna is popular among native-plant enthusiasts and butterfly gardeners in South Florida, but it is little known elsewhere. It is a charming shrub that only reaches about 4' in height, making it a perfect candidate for small gardens or for a planter on a sunny patio. It, along with other species of *Senna* and *Cassia,* is one of the preferred larval host plants of the orange-barred sulphur, cloudless sulphur, and sleepy orange butterflies. There are also a number of butterflies and skippers that visit the bright yellow flowers for nectar.

This species is short-lived and may decline after 3–5 years, but it is fast-growing from seed, so replacing older plants on a rotating basis will ensure that there are plenty around to keep the butterflies happy. A similar and related native species, privet senna (*Senna ligustrina*), is taller and more upright, with narrow, pointed leaflets and paler flowers. These species can be found growing together in the wild, along hammock margins and in pine rocklands of southern Florida. Both are available from nurseries that specialize in Florida native plants.

Saw Palmetto

Serenoa repens
Arecaceae (Palm Family)

Flowering Season: Spring and summer.

Native Range: Southeastern United States.

Comments: Saw palmetto has gained popularity in recent years, not only as a tough, versatile landscape plant but also for pharmaceutical use; an extract from the fruits has gained popularity for helping men maintain a healthy prostate. The fruits are also a favorite food of Florida black bears. Saw palmetto is trouble-free in the landscape, requiring little attention other than occasional removal of dead fronds. In pinelands, the trunks usually creep along the ground, but they are typically upright in garden settings because they are not subjected to fire.

The tiny, fragrant flowers are highly attractive to bees and small butterflies. The rare Atala butterfly is especially fond of the flowers and is a prized butterfly for gardeners in southeastern Florida. Leaves of saw palmetto also host larvae of palmetto skippers.

Saw palmetto looks great when planted with sunny walkways between the clusters and is the perfect plant to create islands of green in the landscape. It grows best in full sun and is unaffected by drought. It also thrives in the sandy, nutrient-poor soils of beach dunes. It has so many virtues it is amazing that it has not become popular until recent years. A silvery-leaved form is also available.

Necklace Pod

Sophora tomentosa
Fabaceae (Pea Family)

Flowering Season: All year.

Native Range: Warm coastal regions of the world.

Comments: Necklace pod is a must-have for hummingbird enthusiasts in tropical Florida. Butterflies also occasionally visit the flowers, and the young leaves serve as larval food for the martial scrub-hairstreak.

Necklace pod produces multiple stems that reach a height of about 6', with long flower spikes topping the stems. The leaves of most cultivated plants of necklace pod in Florida are copiously pubescent (var. *occidentalis*), producing a silvery sheen on the new growth, but this is not the Florida native variety. The leaves of the native variety (var. *truncata*) are only slightly pubescent.

Regardless of which variety you grow, this is a lovely shrub that deserves a sunny spot in tropical Florida gardens. Native-plant enthusiasts and coastal gardeners should seek out the native variety from nurseries that specialize in Florida natives.

It is a grand choice for coastal areas, because it is very tolerant of drought and salt. Seeds are produced in pendent, necklacelike pods. These are poisonous if eaten, so keep this in mind if you have children. If they could pose a health risk to youngsters, be safe by cutting off the spent flowering spikes to prevent fruits from forming.

Caatinga Porterweed

Stachytarpheta caatingensis
Verbenaceae (Verbena Family)

Flowering Season: All year.
Native Range: Bahia, Brazil.
Comments: This species was only recently introduced into the Florida nursery trade by nurseryman Jesse Durko from cuttings of a cultivated plant at Marie Selby Botanical Gardens in Sarasota, Florida. Harry Luther, a bromeliad expert who worked at Selby Gardens, originally collected it in Brazil. It remained unidentified until the author sent a specimen to Drs. Richard Abbott and Ronald Liesner at the Missouri Botanical Garden, who finally put a name on it just in time for this guidebook. Here in our Homestead, Florida, garden the flowers are visited regularly by ruby-throated hummingbirds; of the butterfly visitors, skippers, blues, crescents, and hairstreaks are the most frequent.

Caatinga porterweed reaches 8' tall with a shrubby canopy about half as wide, but it can be kept smaller by pruning if desired. In Brazil, it grows in a semiarid habitat (called *caatinga,* or "white forest," in the Tupi language), so plant it in full sun and keep it on the dry side. It may be some time before it reaches mainstream nurseries, but it is currently sold in specialty nurseries in Miami-Dade and Broward counties. If you live on the west coast, check plant sales at Selby Gardens.

Emerson's Folly

Stachytarpheta 'Emerson's Folly'
Verbenaceae (Verbena Family)

Flowering Season: All year.
Native Range: Of hybrid origin.
Comments: This exciting recent introduction to the Florida nursery trade by Boynton Botanicals in Palm Beach County shows great promise as a premier hummingbird and butterfly attractor for tropical Florida gardens. Owners Mike and Kathleen Kastenholz created it by taking pollen from *Stachytarpheta mutabilis* and using it to pollinate flowers on *S. frantzii*. They named the resulting hybrid 'Emerson's Folly' to memorialize the untimely loss of their son Emerson. The "folly" part of the name refers to the unanticipated change in color as the flowers age from pink to very pale pink with darker highlights. The flower is a fitting tribute, for when the blossoms fade, a heart is formed in the center of each flower. The flowers last for several days each and appear in slow succession on long spikes that terminate each branch tip.

Grow 'Emerson's Folly' in full sun, and it will delight every hummingbird and butterfly within sight. It tops out at about 4' tall, with an equal spread, and it looks especially fashionable in garden settings when intermixed with its two parents. If for no other reason, grow it for Emerson.

Purple Porterweed

Stachytarpheta frantzii
Verbenaceae (Verbena Family)

Flowering Season: All year.

Native Range: West Indies and tropical America.

Comments: Porterweeds are a group of plants that every gardener in Florida should pay close attention to, principally because their small flowers are like magnets for butterflies and hummingbirds. There's even more good news, because some fascinating new hybrids have recently been introduced into the Florida nursery trade.

Purple porterweed is becoming more widely available in nurseries, which can be directly attributed to the growing popularity of butterfly gardening. It reaches 4–5' tall with an equal spread, and it blooms when quite young. It looks stunning when intermixed with pink porterweed (*Stachytarpheta mutabilis*), also treated in this guide. If this sounds appealing to you, then plant them about 5' apart to avoid overcrowding at maturity. These two species are the largest porterweeds grown in Florida, and a hummingbird and butterfly garden in tropical Florida is not complete without them.

Full sun is a must, and light applications of fertilizer in springtime will create a vigorous and healthy plant through the rest of the year.

Blue Porterweed

Stachytarpheta jamaicensis
Verbenaceae (Verbena Family)

Flowering Season: All year.

Native Range: Alabama and Florida through the Bahamas, West Indies, and tropical America.

Comments: There is much confusion regarding Florida's only native porterweed. It has coarsely toothed leaves, and the stems tend to stay low to the ground, seldom reaching more than 20" tall. Small blue flowers are produced throughout the year, and these appear on a thick green stem that somewhat resembles a rat's tail.

Blue porterweed is mandatory for any respectable Florida butterfly garden. It is best used as a low groundcover in dry, sunny locations. The flowers attract a plenitude of butterflies and also the occasional hummingbird, plus the leaves serve as larval food for the tropical buckeye butterfly. It is sold in Florida nurseries, but be aware of an exotic, upright species (*Stachytarpheta cayennensis*) with quilted leaf blades and dark blue flowers. It is weedy, has high seed viability, and can contaminate the gene pool of the Florida native blue porterweed through hybridization.

Another small-growing species is red porterweed (*Stachytarpheta microphylla*), but it is short-lived and cold sensitive. Red porterweed grows to 2' tall and has small red flowers.

Pink Porterweed

Stachytarpheta mutabilis

Verbenaceae (Verbena Family)

Flowering Season: All year.

Native Range: Tropical America.

Comments: The flowers of pink porterweed are an attractive rich coral color when they open; then, over several days, they fade to light pink. It reaches about 5' tall with an equal or greater spread, so if you offer it the room it requires, it can be the center of attention for hummingbirds and butterflies in your garden.

Unlike some species in this genus, the flowers of pink porterweed each last for several days, and there may be two dozen or more blossoms open at once on each flower spike. It is positively dazzling in its prime flowering season and requires little care once established. Its only problem might be twig borers, which can cause stem dieback, but these can be controlled without insecticides by cutting off and discarding affected stems.

It is hardy in Zone 10, but in colder regions it is planted in springtime and grown until it is killed by the first winter frost. It is sometimes available in department store garden centers, so make it a high priority on your shopping list.

Chinese Rain Bell

Strobilanthes cusia
Acanthaceae
(Acanthus Family)

Flowering Season: All year.

Native Range: Asia.

Comments: This quaint shrub grows in moist habitats in its native Asia, so it prefers being watered during dry spells. It produces multiple stems that root easily wherever they touch the ground and may require occasional thinning, which gives you the opportunity to share it with friends.

When in flower it is a real attention getter with its glistening bells of 2" flowers dangling like ornaments on airy, branching stems. Large species of butterflies visit the flowers for nectar while skippers access the sweet liquid by crawling down the throat of the flowers. Hummingbirds perform aerial acrobatics to sip nectar from the pendent blossoms—their antics would make any Cirque du Soleil performer jealous.

In your garden you will want Chinese rain bell to be in full sun or light shade, and in a place where you can water it when necessary. Stems reach 5–6' tall, so take all of this into consideration when selecting a site for it. It is available in South Florida specialty nurseries and from Internet sources, or you can beg for cuttings from someone who grows it. In our garden it adorns an area next to a small stream from a natural swimming pool; it always draws attention from visitors.

Bolivian Petunia

Suessenguthia multisetosa
Acanthaceae (Acanthus Family)

Flowering Season: Spring into fall.
Native Range: Bolivia.
Comments: Bolivian petunia is fast growing and can reach heights of 15' if given support, but it usually only reaches half that height in Florida. The brittle stems tend to arch over if they have no support and will root wherever they touch the ground. Although it is available at specialty nurseries, it is still quite uncommon in Florida nurseries. The 3" flowers appear in eye-catching clusters at the tips of the branches, and a mature specimen is absolutely glamorous in the garden. Besides producing very comely flowers for us gardeners, they are also highly attractive to hummingbirds and butterflies.

In its native Bolivia, it grows along shaded riverbanks, so in cultivation it prefers light shade and moist soil. In full sun and dry soil, the leaves will show signs of stress and the plant will wilt easily. Hard freezes will kill it to the ground, but it resprouts energetically when warm weather returns. It is also called Colombian petunia because it is a popular garden subject there. Hummingbirds are its principal pollinator in the American tropics.

Cape Honeysuckle

Tecoma capensis
Bignoniaceae (Trumpet
Creeper Family)

Synonyms: *Tecomaria capensis.*
Flowering Season: All year.
Native Range: Southern Africa.
Comments: Cape honeysuckle comes with some advisories. It has a strong tendency to spread from root suckers, so these must be kept under control by mowing or pruning them at ground level with shears (never use herbicide on root suckers). It may also spread from seed in cultivation. On the bright side, you can hardly find a more floriferous shrub to add year-round color to your garden, especially along a fence where it will form an excellent privacy screen. The flowers produce a banquet of nectar for hummingbirds and butterflies, plus there are flower colors, ranging from red to yellow, apricot, pink, salmon, and white, from which to choose. There's even a cultivar named 'Rocky Horror' with oversized reddish-orange flowers.

This plant is not related to true honeysuckle. The word "Cape" in its name refers to the Cape of Good Hope at the southernmost tip of Africa. In full sun, it reaches about 8' tall with multiple, often rangy, stems that require seasonal thinning. If you are willing to work at keeping it under control, it is a winner for hummingbirds.

Mexican Sunflower

Tithonia diversifolia
Asteraceae (Aster Family)

Flowering Season: All year.
Native Range: Mexico.
Comments: Mexican sunflower makes a grandiose backdrop for a butterfly garden because the stems reach 8' tall with saucer-sized, 6" heads of cheery yellow sunflowers throughout the year. The small yellow disk flowers are a favorite of many butterflies and occasional hummingbirds, but keep your grub hoe and machete handy because it can claim your entire garden for itself if left unattended for too long.

If you are diligent about pulling up seedlings and keeping stems from rooting when they touch the ground, then you can hardly find a plant that puts on a more spectacular display, and it is great for cut flowers.

It requires full sun and is very drought tolerant once it is established. Propagation is by seed, cuttings, or transplanting rooted stems. Do not become too lackadaisical about keeping Mexican sunflower in bounds, or your neighbors will be listening to swear words they have never heard before. If you have new neighbors and you're looking for a nice house-warming gift, may I suggest Mexican sunflower and a machete?

Tropical Buttercups

Turnera diffusa
Passifloraceae
(Passionflower Family)

Synonyms: *Turnera aphrodisiaca.*
Flowering Season: All year.
Native Range: Texas and Mexico south to Brazil, and from the Bahamas through the West Indies.

Comments: The first time I laid eyes on this quaint shrub in flower was on the island of Abaco in the Bahamas. The stems were covered with dainty, yellow, ½" flowers, and I spent an hour trying to identify all of the butterflies around the blossoms. The next time I saw it was at Fairchild Tropical Botanic Garden in Coral Gables, where I watched Phaon crescent and Cassius blue butterflies poking their tongues down the throats of the flowers. A plant now resides in our home garden.

The older synonym, *Turnera aphrodisiaca*, suggests that it might be worthwhile to brew an herbal tea from the leaves to see what happens. Tropical buttercups is closely related to yellow alder (*Turnera ulmifolia*), but that species is not recommended for home gardening because it escapes into natural areas, especially beach dunes.

Tropical buttercups can reach 6' tall in good growing conditions and requires sun coupled with dry, rocky or sandy soil. Rather than traveling to the Bahamas, Texas, or the Neotropics to collect seeds, your best bet is to keep your eye on plant sales held at botanical gardens. The turnera family (Turneraceae) was recently moved into the passionflower family (Passifloraceae).

Snowflakes

Wrightia antidysenterica
Apocynaceae (Dogbane Family)

Flowering Season: All year.
Native Range: Tropical Asia.
Comments: Snowflakes" and "Arctic snow" seem like odd names for a tropical shrub, but with a bit of imagination I suppose you can envision the flowers resembling flakes of snow. It is also called white angel. Regardless of which name you prefer, the star-shaped, immaculately white, delightfully fragrant flowers adorn the plant throughout the year, attracting hummingbirds and an array of butterflies.

It stays relatively compact at about 3–4' tall, so it is perfect for lining a walkway or being placed on each side of an entryway to greet guests with its perfumed blossoms. It performs best in bright, filtered light or wherever it gets a break from the midday sun of summer. Morning sun followed by transitional shade at midday would be perfect.

It is not the easiest plant to find in Florida nurseries, but if all else fails, seek it out from Internet sources. If you grow it in a container, place it where hummingbirds will have free access to the flowers. If you're a transplant from the North, then maybe the flowers will help remind you of snow. Rum might help.

Vines

This section includes woody and semiwoody vines. Some vines can be very aggressive, but many species are worthy horticultural subjects for humming-bird and butterfly gardens. They can be used to cover fences for privacy, planted on arbors and trellises, or allowed to climb large trees.

Red jade vine, *Camptosema spectabile.*

Carolina Aster

Ampelaster carolinianus
Asteraceae (Aster Family)

Synonyms: *Aster carolinianus; Symphyotrichum carolinianum.*
Flowering Season: Late summer through winter.
Native Range: Southeastern United States.
Comments: It was a difficult decision whether to categorize this species as a shrub or vine, but it is placed here because it has a distinct vining growth habit, sometimes sending long stems 15′ or more up into trees. Because of this habit, it is also called climbing aster. It is a wonderful plant for butterfly gardens, because it blooms through the cool winter months. It prefers moist soil and tolerates sun or shade.

Grown in the open, it can form a mounded shrub with long stems, but it looks more natural to allow it to clamber around on its own just as it does in nature.

The flower heads are about 1¾–2″ across, attracting an array of butterflies. The leaves also serve as larval food for the pearl crescent butterfly. Check the Florida Wildflower Growers Cooperative for seeds (see the Resources section), or check native-plant nurseries. You can also collect your own seeds in late winter and early spring from wild plants, but you need to be timely because a few strong wind gusts will send them flying. There are many more Florida native members of the aster family, so your choices are plentiful.

Giant Pipevine

Aristolochia gigantea
Aristolochiaceae
(Birthwort Family)

Flowering Season: All year.
Native Range: Brazil.
Comments: The bizarre yet bedazzling flowers of this tropical vine measure 6"
across and 10" long. This and other species are larval food of the Polydamas swallowtail, a striking black-and-yellow tailless swallowtail common in the southern
counties. The chocolate-brown larvae aggregate when they are young but eventually become solitary, reaching 2" at maturity.

Some pipevines, such as *Aristolochia grandiflora* and *A. ringens,* produce flowers that smell like putrefying meat, which is a ploy to attract flies as pollinators.
The flowers of giant pipevine have no scent, although the leaves have a peculiar
odor when crushed. Of the *Aristolochia* species in cultivation, this is one of the
best for butterfly gardens, because it is rambunctious enough to sustain a sizable
population of Polydamas swallowtail larvae, and the flowers are a surefire conversation starter. Give it ample room, because it can cover at least 20' of fence in
each direction.

If you would prefer a species with even more outlandish flowers, then try *Aristolochia brasiliensis.* It has mottled brownish-purple flowers that reach 20" long.

Bougainvillea

Bougainvillea glabra
Nyctaginaceae
(Four-O'Clock Family)

Flowering Season: All year.
Native Range: Brazil.
Comments: This popular woody vine has spreading branches and triangular floral bracts that range in color from white, lilac, red, mauve, or purple. There are short thorns that curve at the tip; these are used in the wild to grasp tree branches for climbing, but they also claw the arms of gardeners with pruning shears in their hands. There are numerous named cultivars and hybrids in the trade, so you can choose your favorite color.

The small, tubular flowers are typically white, but most people don't even notice the flowers because of the garishly colored bracts that surround them. Butterflies visit the flowers; they seem to do so more often in some regions than in others, but this may have to do with the availability of other nectar sources. Butterflies commonly visit bougainvillea flowers in the Florida Keys.

Bougainvillea requires support from a fence or a strong arbor, or be allowed to scramble into a large tree. The latter option is ill-advised, however, because it creates wind resistance in the tree canopy during strong tropical storms. If you like bougainvillea, then by all means plant one. Because of its high maintenance and sharp thorns, my personal preference is to admire it in someone else's yard. 🦋

Red Jade Vine

Camptosema spectabile
Fabaceae (Pea Family)

Synonyms: *Camptosema grandiflorum.*
Flowering Season: Winter.
Native Range: Brazil.
Comments: Gardeners often overlook vines when it comes to attracting hummingbirds, but there are many species that become their center of attention when in flower. Red jade vine is one of them. Each winter, cascading spikes of yawning red flowers appear on this twining vine, and hummingbirds practically stand in line to get at the tempting blooms.

You will need to construct a tall pergola so you can walk beneath it to admire the pendent floral spikes created by this ornate Brazilian vine. We have it growing on a pergola along with the real jade vine (*Strongylodon macrobotrys*), which produces eye-popping, pendent spikes of aquamarine flowers in April and May. Too bad they don't bloom at the same time.

Do not mistake *Camptosema spectabile* with the related red-flowered species *Mucuna bennettii,* or scarlet jade vine. It has red, pealike flowers on pendent spikes. Both are grown in Florida by collectors, but *Camptosema* is more cold tolerant and is much preferred by hummingbirds. If you are unsuccessful at locating a source locally, red jade vine is widely available on the Internet.

Spurred Butterfly Pea

Centrosema virginianum
Fabaceae (Pea Family)

Flowering Season: All year.

Native Range: Southeastern United States and tropical America.

Comments: The neat thing about this Florida native vine is that it needs very little space and is quite content to twine unobtrusively through shrubs or simply trail across open ground in full sun. In its native habitat, it is adapted to fire and will bloom in profusion just weeks after a burn has reduced competition and exposed the soil to direct sunlight.

As pretty as the plant is when in flower, it remains a seldom-cultivated plant and is only grown by native-plant enthusiasts. While many vines climb by tendrils or hooked thorns, this vine loosely twines or merely lounges on whatever support is nearby. The 1½" flowers may be pale violet, light pink, or rich purplish violet. The fruits are flat pods, so look for them if you are planning to grow it from seed. It is very common in pineland habitat.

Besides its colorful blossoms, its interest to gardeners is also as a preferential larval host plant of the long-tailed skipper and the Dorantes longtail butterflies. Both are common in urban gardens. A key behavioral difference between these look-alike butterflies is that the Dorantes longtail generally keeps its wings closed when it lands.

Java Glory Vine

Clerodendrum × *speciosum*
Lamiaceae (Mint Family)

Flowering Season: All year.

Native Range: Of hybrid origin.

Comments: Java glory vine is a hybrid between *Clerodendrum thomsoniae* (treated next) and *Clerodendrum splendens*. The latter is a vining species that produces showy clusters of bright red flowers around Christmastime. This hybrid is everblooming with red flowers subtended by violet calyces. The flowers are frequented by nectar-seeking hummingbirds and butterflies.

It will require a fence, arbor, or tall trellis to twine upon, and you will need to be an astute gardener to control seedlings and root suckers. Root suckers can be mowed, but if this is not an option then use pruning shears to lop them off at ground level. If planted on a fence, it will grow dense enough to provide a pretty privacy screen.

It is readily available even in department store garden centers. Bleeding hearts (*C. thomsoniae*) is much less aggressive in garden settings but is not nearly as floriferous. Planting the two together so that they intertwine creates an elegant display of violet and white calyces combined with red flowers. You can grow either one in a large container with a trellis if you are good at pruning.

Bleeding Hearts
Clerodendrum thomsoniae
Lamiaceae (Mint Family)

Flowering Season: All year.

Native Range: Africa.

Comments: This adorable vine can be grown on a trellis, an arbor, or a fence, or be allowed to twine up small trees. Prune back the vine severely if it outgrows its space, because pruning during late winter will encourage new flowering stems come spring. Stems may reach 16' long if not pruned. Like the preceding species, if you do not have room for a rambling vine in your garden, then keep it in a large pot and let it climb a trellis.

Brilliant scarlet flowers emerge from snow-white calyces, and the contrast between these two colors is stunning. Hummingbirds and a host of butterflies crowd around a flowering plant.

Although it fell out of vogue for years, it seems to be making a comeback due, perhaps, to the growing popularity of butterfly gardening. It is hardy in Zone 10 but can be grown farther north by cutting the stems off in the fall and covering the roots with mulch for winter protection.

Not enough can be said to describe the beauty of this vine, so I highly recommend that you find a sunny spot for it in your garden. The hummingbirds and butterflies will appreciate it, too.

Bluepea

Clitoria ternatea

Fabaceae (Pea Family)

Flowering Season: All year.

Native Range: Old World tropics.

Comments: There is hardly a more regal color in nature than the blue flowers of this petite vine. You have a choice between the typical single-flowered form (pictured) and an alluring double form. Both are widely cultivated in Florida. The genus name alludes to the keel of the flower, named for its resemblance to female genitalia, in case you'd like something interesting to explain to guests during garden tours. The species name refers to the island of Ternate in the northern Moluccas, a chain of Indonesian islands most famous for the production of cloves.

The leaves are larval food for the long-tailed skipper and the Dorantes longtail. Bluepea is easy to grow and is petite enough to plant on a small trellis or be allowed to ramble around on a fence with other vines.

It produces pods readily and may spread from seed around your garden, so pulling up seedlings may be a seasonal chore; it might be easier to remove the pods before they mature. If for some bizarre reason you would like to serve blue rice for dinner, you can create it by adding some flowers to the cooking water.

Orange Flame Vine

Combretum fruticosum
Combretaceae
(Combretum Family)

Flowering Season: Summer into fall.

Native Range: South America.

Comments: In flower, this exuberant vine will command attention. The upward-pointing stamens resemble long-bristled brushes, opening yellow then turning orange as they age. Each arrangement of flowers is about 4–6" long, and blooms keep coming for many weeks. Swallowtails and other large butterflies visit the flowers for nectar, and if the flowers continue into the fall, hummingbirds will visit the blossoms as they arrive at their winter home in southern Florida.

Orange flame vine will need strong support from a tall arbor or a fence. Although it is a tropical species, it has demonstrated a tolerance to temperatures well below freezing, so it is suitable to grow north of Zone 10 in Florida.

It may take some detective work to find this vine in local nurseries, but be sure to attend plant sales in botanical gardens, especially in South Florida. If that fails, then fear not, for it is abundantly available from Internet sources. It requires full sun to produce its lavish blooms.

Eastern Milkpea

Galactia volubilis

Fabaceae (Pea Family)

Flowering Season: All year.

Native Range: Eastern United States.

Comments: This species is widespread throughout Florida, but do not expect it to be available in nurseries unless you happen to find a local Florida grower who specializes in native plants and likes to grow a little bit of everything. You will more than likely have to find a seed source from wild plants, but that will not be difficult, because this common species can be found along mowed road shoulders, edges of forests, and along fencerows. It flowers and sets pods all year, so a little exploring should pay off.

Let this vine twine through shrubs somewhere in the back of your garden, because it will likely have chewed leaves on a regular basis: it is the larval host plant of the Cassius blue, silver-spotted skipper, long-tailed skipper, and zarucco duskywing. For such a petite little vine, it offers a lot to butterfly gardeners.

The vine produces small but pretty pink flowers year-round, which are followed by beanlike pods. Once it is established, it will spread from seed on its own, so remove or transplant seedlings as necessary. The related Elliott's milkpea (*Galactia elliottii*) has larger white flowers and is the larval host plant of the zarucco duskywing.

Lady Doorly's Morning-Glory

Ipomoea horsfalliae
Convolvulaceae
(Morning-Glory Family)

Flowering Season: Mostly winter.

Native Range: West Indies to Brazil.

Comments: If you're looking for a fabulous vine to cover a fence, arbor, or trellis, then consider Lady Doorly's morning-glory. The 2" trumpet-shaped blossoms are hot rosy magenta, and a well-grown vine can be covered with many dozens of eye-popping flowers in winter. Skipper butterflies crawl down the throat of the flowers to access the bountiful nectar, as do honeybees. Ruby-throated hummingbirds also load up on nectar from the flowers.

A single vine can cover at least 20' of fence, so give this species ample room to twine and ramble. The vine is dense enough to create a privacy screen on a fence, or you can simply install a freestanding section of fence elsewhere on your property and plant the vine on it, which would work nicely as a screen for a hidden garden. It would also look stunning on a tall arbor over an entranceway, so choose your best option.

For the longest time this species was only found in botanical gardens, so it's exciting that it is now available in a few specialty nurseries in Miami-Dade County. There are Internet nurseries that sell it as well.

Man-in-the-Ground

Ipomoea microdactyla
Convolvulaceae
(Morning-Glory Family)

Flowering Season: Spring into fall.

Native Range: Southern Florida (Miami-Dade County), Bahamas, and Cuba.

Comments: This Florida native is a state-listed endangered species and occurs in pine rockland habitat in southern Miami-Dade County, including Long Pine Key in Everglades National Park. The common name relates to the underground tuberous root that resembles that of the related sweet potato (*Ipomoea batatas*), which helps the plant survive drought and fire by storing nutrients.

This species is an excellent choice for a trellis or small fence, or you can allow it to twine across shrubs or small trees as it does in pinelands. It will not create a privacy screen as efficiently as other vines, but the sensational pink to pinkish-red, trumpet-shaped flowers makes it an excellent choice for a native garden. It is relatively short-lived and may need to be replaced every few years.

Skipper butterflies and occasional hummingbirds visit the flowers and sphinx moths visit them after sunset. It may take some diligence to locate a source for this vine, but check nurseries in Miami-Dade County that specialize in Florida native plants. Another excellent choice is the native tievine (*Ipomoea cordatotriloba*), which has purple or pale pink flowers.

Coral Honeysuckle

Lonicera sempervirens
Caprifoliaceae
(Honeysuckle Family)

Flowering Season: Spring to fall.

Native Range: Eastern United States south through Central Florida.

Comments: Coral honeysuckle blooms most abundantly in springtime, just before hummingbirds are about to leave South Florida on their migratory journey northward. It is popular in gardens from Central Florida northward but will grow just fine in gardens throughout the state. After it finishes its grand springtime show, it blooms off and on into fall.

Although it is a vine, if you want it to grow into the canopy of a tree or a tall support, it will have to be trained; most gardeners give it a low support and let it form a twining shrublike plant. It is a perennial favorite of hummingbirds. Large butterflies like swallowtails also visit the long, tubular blossoms.

It is widely available and easy to grow in full sun to partial shade. Prune it back in winter to keep it more manageable, but it is called "well behaved" by most gardeners.

Cultivars include a yellow-flowered form called 'John Clayton' and a blazing red form called 'Major Wheeler.'

Climbing Hempvine

Mikania scandens
Asteraceae (Aster Family)

Flowering Season: All year.
Native Range: Eastern United States.
Comments: Pay attention, butterfly lovers. This twining vine occurs throughout Florida, and when a mature specimen is in flower, it is engulfed with clusters of small, white, highly perfumed flowers, which are in turn covered with butterflies. Astute gardeners who are aware of its butterfly-attracting attributes are the only people who cultivate it. It is not commercially available, although there may be some small local nurseries that grow it for native-plant aficionados.

In its wild habitat, it scrambles and twines over other vegetation and can reach the canopy of small trees. It occurs in a variety of habitats, often in and around freshwater wetlands, but also along forest margins. You will likely be on your own to collect seeds, but they sprout easily and the vine grows rapidly. Plant it on a fence, trellis, or where it can clamber around on other vegetation as it does in nature.

A list of butterflies that visit the flowers would literally be a list of butterflies that occur within its range.

Passionflowers

Passiflora spp.
Passifloraceae
(Passionflower Family)

Flowering Season: All year.

Native Range: Mostly the Americas but also Asia, Africa, and Australia.

Comments: Passionflowers are mostly vines that produce lavish and often seductively perfumed flowers. About 500 species are known, and 95 percent of them are indigenous to South America. To add to the menagerie, there are more than 300 hybrids and cultivars, so your choices are abundant.

The larvae of the zebra longwing, Julia, Gulf fritillary, and variegated fritillary all feed on the leaves of many, but not all, passionflowers. For instance, the red-flowered *Passiflora coccinea* and *P. vitifolia* are not utilized, nor is the showy, super-fragrant *Passiflora alata*. And the same goes for *Passiflora edulis*, the passion fruit of commerce. Passionflowers globally are the almost exclusive hosts for more than 70 species of butterflies, but some develop egg-mimicking glands on the petioles in an apparent attempt at deception to keep butterflies from laying real eggs.

Passionflowers will not only add striking beauty and fragrance to your garden but also ensure that you have a thriving population of heliconid butterflies around for your viewing pleasure. The hybrid pictured was developed in Chicago in 1991 and is called 'Lady Margaret.'

Maypop

Passiflora incarnata
Passifloraceae
(Passionflower Family)

Flowering Season: Spring to fall.

Native Range: Pennsylvania to Kansas south through Texas and most of mainland Florida.

Comments: Maypop is Florida's showiest native passionflower, with extravagant flowers that emit an enticing fragrance. It is also called apricot vine and purple passionflower. The Cherokee in Tennessee called the vine *ocoee* and this name was later used for a Florida town and an Appalachian river. It is a larval host plant of the zebra longwing, Julia, Gulf fritillary, and variegated fritillary butterflies in wild habitats and in gardens. Adult butterflies also sip nectar from the blossoms of most passionflowers.

This is an aggressive vine with stems to 20' long or more, and it usually produces root suckers far from the main stem. It is best grown on a fence or arbor, where root suckers can be mowed to keep them in check. It is decidedly cold tolerant, so no winter protection is necessary in Florida.

A beautiful hybrid called *Passiflora* 'Incense' produces exquisitely frilly purple blossoms with a sweet, spicy perfume. It is a cross between *Passiflora incarnata* and *P. cincinnata*. 🌿 🦋 🐛

Corkystem Passionflower

Passiflora suberosa
Passifloraceae
(Passionflower Family)

Flowering Season: All year.

Native Range: Florida, West Indies, and tropical America.

Comments: This is Florida's most common native passionflower, but it often goes unnoticed due to its small stature. It is a preferred larval host plant of the zebra longwing, Julia, Gulf fritillary, and variegated fritillary butterflies. In the landscape it is best to scatter a number of plants around, so they can twine into shrubs. Hopefully one or two will avoid detection and be able to set seed. Like all passionflower species, it produces tendrils that wrap around twigs and other supports, but it can be grown in a hanging basket and allowed to trail over the sides. It can also be used as a groundcover if planted in the open, although this will expose the butterfly eggs and larvae to easier predation by ants.

The green to greenish-white, dime-sized flowers are difficult to see without close scrutiny. Leaf shape is extremely variable, and it is said that the plant produces such variable leaf designs to avoid being detected by female butterflies looking to lay eggs. If true, it sure doesn't work very well. If you grow only one plant, it will be consumed in due haste.

Wild Allamanda

Pentalinon luteum

Apocynaceae

(Dogbane Family)

Flowering Season: Spring through fall.

Native Range: Florida and the West Indies.

Comments: If you take a leisurely drive from spring through fall along the main road in Everglades National Park, once you're about halfway to Flamingo, you will begin seeing beautiful vines twining around the roadside vegetation showing off trumpetlike yellow blossoms. Stop and you will find that the flowers belong to this Florida native vine. It is so showy that it has even received attention from the Florida nursery trade.

You can hardly find a prettier and more floriferous native vine, and it is the perfect candidate for an arbor arching over an entranceway. It is also suitable for a fence or being allowed to twine through shrubs as it does in the Everglades. Skipper butterflies cannot stay away from the flowers, and they crawl down the floral tube to access nectar. This is a relative of the commonly cultivated *Allamanda cathartica* native to the American tropics, but wild allamanda is better behaved in gardens.

Give it a sunny spot and something to climb on, and the only thing left to do is enjoy its blossoms and watch skippers come and go all day long. Remember that most members of this plant family are toxic to eat, and the sap can be irritating to the skin, eyes, and mouth.

Pink Trumpet Vine

Podranea ricasoliana
Bignoniaceae (Trumpet
Creeper Family)

Flowering Season: Late spring into midwinter.

Native Range: South Africa.

Comments: If you want a stunning privacy screen for a chain-link fence, with the added bonus of making every hummingbird in your neighborhood ecstatic, here's your chance. This tropical vine produces glamorous, trumpet-shaped flowers throughout the warm months, and the showstopping floral extravaganza lasts well into winter. All it requires is support in the form of a fence or arbor in full sun or high transitional shade. If you do not have a form of support, then it can be maintained as a shrub through selective pruning. If you are a creative gardener, try growing it espaliered on a wall.

It will benefit from a layer of mulch over the root zone and light applications of fertilizer at the beginning of the rainy season. Besides hummingbirds, skipper butterflies like to crawl down the throat of the flowers to sip nectar.

It is hardy to Zone 9A, so no winter protection is necessary in tropical Florida. Look for it in local retail nurseries and also in department store garden centers.

White Twinevine

Sarcostemma clausum
Apocynaceae
(Dogbane Family)

Flowering Season: Summer through winter.

Native Range: Central and southern Florida, West Indies, and tropical America.

Comments: If you want to add this native vine to your garden, you must collect your own seeds, because you will not find it in nurseries. It is even surprisingly rare to find in gardens designed to attract butterflies. White twinevine is common in mangrove forests, maritime hammocks, and cypress swamps where it twines on shrubs and tree branches. The milky sap from broken stems or leaves may cause a burning sensation if it comes in contact with your eyes or mouth, so wash your hands if you touch the sap. This is a larval host plant of soldier and queen butterflies and occasional monarchs, plus the flowers attract adult butterflies for nectar.

The elongated pods split open to reveal flattened seeds attached to silky floss, so collect them before the wind sends them soaring.

In the garden, this vine can be allowed to twine on shrubs or along a fence. It is rather leggy and may attract aphids, but spraying a 10 percent liquid detergent and water mixture, then hosing it off, will control them. You can also purchase and release lady beetles, to the delight of children. See the Resources section under Organic Pest Controls.

Hawaiian Sunset

Stictocardia beraviensis
Convolvulaceae
(Morning-Glory Family)

Flowering Season: Periodically all year.
Native Range: Tropical Africa.
Comments: You have not seen "pretty" until you've laid eyes on the flowers of this elegant climber. It is dazzling in flower, and the lush, heart-shaped, velvety leaves are beautiful too, adding a tropical flair to any garden setting. The fragrant flowers are 2–3" wide and last for several days each. Hummingbirds visit them for nectar and pollen, and skipper butterflies crawl inside the corolla for nectar.

One drawback is that this fast-growing vine requires substantial support in the form of a fence, an arbor, or a very stout trellis. In a few short months, it will entirely engulf anything it is planted on. If you have a very large tree on your property, it could be trained to climb into its canopy, but this will place the flowers too high to enjoy. You can try growing it in a container next to a tall trellis, but you will need a 7- to 10-gallon pot size.

Although Hawaiian sunset will flower in shady situations, the rule of thumb is the more sun, the more flowers. Give this vine plenty of water, and apply fertilizer in springtime. You will find it at specialty nurseries and through Internet sources.

Black-Eyed Susan Vine

Thunbergia alata
Acanthaceae (Acanthus Family)

Flowering Season: All year.
Native Range: Africa.
Comments: The eye-popping flowers of this African vine are sure to catch the attention of your gardening friends, along with that of hummingbirds and a wide variety of butterflies, especially swallowtails and skippers. If you are not partial to orange, there are other varieties with red, reddish-orange, pale yellow, bright yellow, and even white blossoms, and most are further adorned with the dark chocolate-purple center that tells butterflies and hummingbirds where they should stick their tongues. A hybrid called 'Blushing Susie' sports red flowers mixed in with shades of ivory and apricot.

The stems may reach 8' long, so it will require some means of support. It is a perfect vine for an arbor over an entranceway and can be pruned in springtime to encourage new stems and more flowers.

The orange and bright yellow varieties are widely cultivated, but locating some of the more unusual color forms will take some effort on your part, so check Internet sources. Try planting several different color forms together on a fence, and enjoy the riot of color.

Maiden's Jealousy
Tristellateia australasiae
Malpighiaceae
(Malpighia Family)

Flowering Season: Periodically throughout the year.

Native Range: Southeast Asia to Australia and New Caledonia.

Comments: If you like cheerful, bright yellow blossoms, you will be tickled pink with this small-growing tropical vine. Each inflorescence may produce as many as 20 or 30 red-stamened yellow flowers that are attractive to both hummingbirds and butterflies.

Although it is vigorous, it is small enough to be grown on a trellis or a small arbor. In Great Britain it is a popular indoor plant. Other common names in the nursery trade are Australian gold vine, shower of gold, and thryallis vine. It blooms off and on throughout the year, but is at its prime from July through November. It can be grown in full sun to light shade and tolerates winter temperatures down to about 50°F. Cover it with a blanket or bring it indoors if colder weather is expected. If it does get killed back, it will resprout with vigorous new growth in spring.

Availability is limited in Florida nurseries, but it is widely available through many Internet sources. The common name relates to maidens being jealous of its beauty.

Cowpea

Vigna luteola

Fabaceae (Pea Family)

Flowering Season: All year.

Native Range: Southern United States, Bermuda, and tropical America.

Comments: You will need to look along roadsides and canal banks for this weedy vine and collect seeds yourself, because even nurseries that specialize in Florida native plants do not cultivate it. It is commonly found in disturbed sites scrambling around with other weedy species and can be especially common along roads that are adjacent to farm fields. The seeds are in hairy pods.

If you are wondering why you should grow a roadside weed, it is because one person's weed is another person's butterfly attractor. It is a favored larval host plant of the gray hairstreak, Cassius blue, Ceraunus blue, Dorantes longtail, and long-tailed skipper butterflies. The last four species are relatively common urban butterflies; the gray hairstreak is locally and seasonally common.

Allow cowpea to scramble around on other plants just as it does along roadsides. It will probably not have the propensity to become a troublesome weed in your garden, but unwanted seedlings can be removed when necessary if butterfly larvae do not keep it under control.

Herbs, Subshrubs, and Groundcovers

This section includes plants without woody stems as well as herbaceous and semiwoody subshrubs and groundcovers. Species included here are very versatile in the landscape.

Southeastern sunflower, *Helianthus agrestis*.

Lipstick Plant

Aeschynanthus × *splendidus*
Gesneriaceae (Gesneriad Family)

Flowering Season: Summer into early winter.

Native Range: Of hybrid origin.

Comments: This glamorous plant is a hybrid between *Aeschynanthus speciosus* and *A. parasiticus,* both epiphytic in tropical rainforests of Southeast Asia, where the stems hang down from tree branches. Sunbirds, Asia and Africa's long-billed counterparts to hummingbirds, pollinate the flowers in the wild.

Grow lipstick plant in a hanging wire basket that is lined with fiber matting and filled with a mixture of peat and perlite. Replenish the spent soil mix at the end of the growing season, and keep it watered in the dry season. It prefers filtered light, so hang the basket from a low tree branch where it receives morning sun with filtered light the rest of the day. Drench your plant in spring with liquid fertilizer that has been diluted from its recommended strength by half.

Ruby-throated hummingbirds delight in the flowers, and giant swallowtail butterflies visit them as well. If you cannot find lipstick plant locally, it is readily available through Internet sources. Some color forms produce bright yellow or uniformly red flowers. All of them are absolutely stunning.

Mexican Balm

Agastache mexicana
Lamiaceae (Mint Family)

Flowering Season: All year.
Native Range: Mexico.
Comments: Gardeners have quite a few choices of species, varieties, cultivars, and hybrids of the genus *Agastache*; they come in every color of the rainbow. Mexican balm is just one of the possibilities. The color form pictured is 'Acapulco Orange.' From Texas to California, this species is one of the top choices of butterfly and hummingbird gardeners, and it performs very well in Florida if its horticultural needs are met.

It requires full sun and alkaline soil that is amended with perlite or coarse gravel so the roots remain aerated; otherwise, high humidity may cause overall decline. Try growing it in a container filled with cactus soil amended with perlite or coarse gravel.

The leaves are aromatic and make a pleasant, lemon-flavored herbal tea. They can also be added to soups and other dishes for a lemony accent. A hybrid between two other species, 'Blue Fortune,' has small, butterfly-attracting blue flowers and is also grown as a culinary herb. The most difficult choice you will have with this genus is deciding which colors to choose. Check Internet sources.

Garlic Chives

Allium tuberosum
Amaryllidaceae
(Amaryllis Family)

Flowering Season: Summer into winter.

Native Range: Tropical Asia.

Comments: It is nice when the stars align and you find a useful plant for both you and the butterflies, and this is what you get with garlic chives. The leaves are widely used in Asian cooking and impart a garlic flavor to stews, soups, and stir-fried dishes, plus butterflies simply cannot resist a patch of it in flower.

Garlic chives have clusters of white flowers on tall stalks that stand well above the leaves. Onion chives (*Allium schoenoprasum*) produce rounded clusters of violet flowers. Both species can be commonly found in garden centers where potherbs are sold. Plant enough so there will be plenty of leaves for you to harvest, allowing an abundance of flowers for the butterflies.

Queen and monarch butterflies are particularly partial to the flowers, as are bees. Choose a sunny spot in your garden, or install a raised bed as a dedicated herb garden right outside your kitchen for easy access. Clumps of garlic chives keep expanding, and their leaves remain all year in warm regions but will die back in cold temperate zones during the winter. Keep it contained, because it is a pernicious weed in some countries. As a butterfly attractor, it gets two thumbs up.

Dill

Anethum graveolens
Apiaceae (Celery Family)

Flowering Season: All year.

Native Range: Mediterranean to Asia Minor.

Comments: This is the familiar culinary herb whose seeds are used to pickle foods; its leaves impart a distinctive flavor to salmon and many other dishes. Its tiny, fragrant flowers are in clusters that resemble yellow fireworks; these attract small butterflies and diurnal moths. Also, the leaves are a favorite larval food for the black swallowtail butterfly.

Dill is an annual, so plant seeds periodically throughout the year, especially in fall. In tropical Florida, black swallowtails will generally be looking for their larval host plants in winter and spring, but butterflies will visit the flowers any time of year they are available.

If you have a dedicated herb garden, then plant a row of dill, fennel, and parsley, and you will have plenty of herbs to flavor dishes before black swallowtail larvae help consume them. There are other, Florida native members of the celery family that also attract black swallowtails, but they are mostly freshwater wetland species: water dropwort (*Oxypolis filiformis*) and water hemlock (*Cicuta maculata*). Be advised, however, that water hemlock is the most poisonous native plant in Florida.

Scarlet Milkweed

Asclepias curassavica
Apocynaceae
(Dogbane Family)

Flowering Season: All year.
Native Range: Tropical America.
Comments: Of all the milkweeds in the Florida nursery trade, this is the most reliable species for attracting monarch, queen, and soldier butterflies. It is a favored larval host plant, and adult butterflies drink nectar from the flowers, too. The plants can be entirely defoliated by hungry larvae; then they resprout from the stems, only to be defoliated again.

Flowers of this species are typically two-toned with reddish orange and yellow, but there is also a pure yellow-flowered form available. Scarlet milkweed spreads readily on its own from seed, but if you choose to propagate milkweed seeds in pots or flats, make certain you cover the seedlings with screen—or you may find your entire crop eaten by a single larva.

This is a popular plant and is commonly available. It will reach about 3' tall. Milkweeds contain heart toxins (cardiac glycosides), and the milky sap may cause mild dermatitis on sensitive people. If you're one to keep up with taxonomy, the entire former milkweed family (Asclepiadaceae) has recently been relegated into the dogbane family (Apocynaceae). Another non-native species to consider is *Gomphocarpus physocarpus,* which reaches 6' tall and bears pendent clusters of white flowers and balloonlike pods. 🦋 🐛 ☠

Rose Milkweed

Asclepias incarnata

Apocynaceae

(Dogbane Family)

Flowering Season: Spring through summer.

Native Range: Indiana, Illinois, and Missouri south into Central Florida.

Comments: Rose milkweed inhabits freshwater wetlands, and its natural range extends south in Florida to the Fakahatchee Swamp in Collier County. The pink flowers create an attractive display, but you must provide this plant with reliably wet soil by using automatic irrigation or by setting a pot of plants in a dish of water.

If you can locate an old abandoned bathtub, and you like tacky yard art, place it in a sunny location, plug the drain hole, and then fill it full of soil. Next, simply run the garden hose in the tub until water reaches the top: you have now become the proud owner of a freshwater wetland for your swamp-loving plants. Please do not tell your neighbors or code-enforcement officers where you received this treasured information.

Rose milkweed is available from specialty nurseries within its natural Florida range, and you can find plants and seeds offered through many Internet sources.

The larvae of monarch and queen butterflies relish this species, so expect it to be seasonally devoured. Other water-loving native species to consider are swamp milkweed (*Asclepias perennis*), which has clusters of tiny pale pink flowers, and prairie milkweed (*Asclepias lanceolata*), whose blossoms are two-toned with red and yellow.

Butterfly Milkweed

Asclepias tuberosa
Apocynaceae
(Dogbane Family)

Flowering Season: Summer through fall.

Native Range: Eastern United States through the Midwest and south into northern Mexico.

Comments: This Florida native is common throughout much of mainland Florida and is sold by nurseries that specialize in native plants. It is also available in seed packets and is superb for a dry, sunny location where butterflies have access to the flowers for nectar. Monarch and queen butterfly larvae feed on the flowers and fresh young leaves, while the hairy, colorful larvae of the echo moth (pictured) might consume the entire plant.

The flowers range from pale orange to deep reddish orange and are produced in dense, flat-topped clusters. It is quite ornate when in flower and looks best when planted in beds. It prefers sandy soil that drains quickly and is decidedly drought tolerant, making it a perfect candidate for xeriscaping. It is also perfectly adaptable for a container placed in a sunny location. In good conditions, the plant may reach about 18" tall.

Milkweeds contain toxins that make butterfly larvae poisonous and distasteful to birds and mammals. The genus *Asclepias* was named to honor Aesculapius, the ancient Greek god of medicine and healing. It is also called pleurisy root because the roots of this species were once used to treat pleurisy, a respiratory illness.

Herb-of-Grace

Bacopa monnieri
Plantaginaceae
(Plantain Family)

Flowering Season: All year.

Native Range: Southeastern United States, West Indies, and British Honduras.

Comments: Why this plant is not more widely cultivated by gardeners in Florida is puzzling. It will survive growing in lawns that are overwatered (which is usually the case), or it can be grown in a wide, shallow container where the stems can cascade over the sides. You can also plant it along a patio border, where it can receive regular watering. The stems are succulent, so it will not tolerate continuous foot traffic.

It is a worthy garden subject because of its quaint, pinkish flowers, but it is also a preferred larval host plant of the white peacock butterfly. Among the nectar gleaners are many small, low-flying butterflies, including the white peacock. This common urban butterfly has wings colorfully marked with white, orange, brown, and black.

Herb-of-grace is a low, prostrate species commonly found in freshwater wetlands throughout Florida, so look for it along roadsides that bisect its natural habitat. Although it can be grown from seed, the stems root wherever they touch ground, so divisions can be transplanted with ease. It is not a species you will easily find in cultivation, but check nurseries that specialize in native plants.

Spanish Needles

Bidens alba var. *radiata*
Asteraceae (Aster Family)

Flowering Season: All year.

Native Range: Tropics and subtropics of both hemispheres.

Comments: No introduction is necessary for this Florida native, which exemplifies Ralph Waldo Emerson's observation that a weed is "a plant whose virtues have not yet been discovered." Those who are aware of its attractiveness to butterflies are the only souls who tolerate it in gardens. Spanish needles not only is exceptional at attracting butterflies but also is the larval food for the dainty sulphur butterfly.

The best way to treat this plant in gardens is to allow it to remain in places where it is not a nuisance, and then either remove it, cut it back, or mow it when it becomes too rangy. The two-pronged, sticklike seeds cling to fur, feathers, and clothing, and this is how you, your pets, and other mammals spread it around.

Find a patch of Spanish needles, and you will likely find plenty of butterflies, so consider this when you are weeding your garden. Yes, it meets every definition of a weed, but it performs better than most plants at attracting butterflies. The clear juice from broken stems will stop small cuts from bleeding almost instantly, and the young leaves contain 50 percent more iron than spinach, if that helps you appreciate it a little more.

Yellow Canna
Canna flaccida
Cannaceae (Canna Family)

Flowering Season: Mostly spring through summer.

Native Range: Southeastern United States.

Comments: Few other plants brighten roadside ditches and wet meadows in Florida as much as our native yellow canna. Each flower is short-lived, but they appear in a cycle for many weeks on each plant. Cannas spread by underground rhizomes, so it is easy to divide plants and move them to other locations in your garden.

Throughout Florida, it is a larval host of the Brazilian skipper butterfly, and the voracious larvae can consume all of the leaves and even chew the stems down to the ground. Cannas once became all the rage for gardeners in the eastern United States, causing Brazilian skippers to spread well north of their historical range. When gardeners tired of having their plantings continuously devoured, cannas slowly fell out of vogue and Brazilian skippers retreated back to their historic range where the native yellow canna grows. The larvae also use the native fireflag (*Thalia geniculata*) found in freshwater wetlands.

Grow yellow canna in a moist sunny spot, and enjoy the flowers during times when the Brazilian skippers haven't discovered them yet. Look for it in nurseries that grow Florida native plants, or collect seeds of plants in rural roadside ditches. When in flower it is impossible to miss.

Indian Shot

Canna indica
Cannaceae (Canna Family)

Flowering Season: All year.
Native Range: India.
Comments: Indian shot flowers are typically red, but there are some interesting bicolored cultivars (pictured) in the trade. This species spreads from seeds and rhizomes to the point of being a pest, but it isn't often when you find a plant that attracts hummingbirds and adult butterflies plus serves as a larval host plant. We gardeners are easily persuaded.

Brazilian skipper larvae make two parallel slits in the leaves and then fold the flap over, holding it together with silk in order to hide from predators during the day, but orioles have learned to pull the flaps open to reveal the morsel hiding inside.

Indian shot reaches 5' tall, but flower spikes can tower above a person's head. The leaves may be green or tinged around the edges with reddish purple.

Indian shot is not widely available in nurseries, but check Internet sources. Or you could find someone who grows it and knock on their door with a shovel in your hand. They may even pay you to take some.

The name "Indian shot" relates to Native Americans using the hard round seeds to load shotgun shells in lieu of lead shot for bird hunting.

Florida Paintbrush
Carphephorus corymbosus
Asteraceae (Aster Family)

Flowering Season: Midsummer to fall.

Native Range: Florida, Georgia, and South Carolina.

Comments: This native wildflower is frequent in upland sandy habitats throughout much of the Florida peninsula into the Panhandle. It does not occur naturally in Everglades National Park, but it does occur in the Corkscrew Swamp and CREW Marsh region of Collier County and from Broward County on the east coast northward.

Butterflies frequent the flowers of Florida paintbrush, so if you have a native wildflower garden, or want to start one, be certain you include this beautiful species. It is not widely available so check nurseries that specialize in Florida native wildflowers or watch for plant sales sponsored by your local Florida Native Plant Society chapter.

Other related species to consider include vanilla leaf (*Carphephorus odoratissimus*), hairy chaffhead (*Carphephorus paniculatus*), and pineland chaffhead (*Carphephorus carnosus*). All of these live in wetter areas than Florida paintbrush, and all are superb butterfly attractors. Florida paintbrush will require a sandy location in full sun; if such conditions do not occur naturally on your property, then create them by making a raised bed filled with a sand-based soil mix.

Silver Cock's Comb

Celosia argentea
Amaranthaceae
(Amaranth Family)

Flowering Season: Spring into winter.
Native Range: India.
Comments: This is a fast-growing annual or short-lived perennial that may reach heights of 6' or more. The leaves are long and slender with each stem topped by a dense spike of brilliant magenta flowers. The bracts turn silver as they age and dry. The flowers attract virtually every skipper butterfly that comes near them, and they occasionally draw hummingbirds. Painted buntings and sparrows savor the small black seeds, which are produced in abundance.

Plant it in a bed, or allow it to overtop shorter plants in the garden. Nurseries that specialize in flowering plants or butterfly-attracting plants may have plants for sale in spring or early summer.

This species performs best in full sun and well-drained soil, and seeds will germinate year after year in areas where there is exposed soil. If you do not have areas like this, then collect seeds each year and germinate them in pots. It does have weedy tendencies, so this is one of those plants that you'll be able to share with your neighbors, whether they want it or not.

Purple Thistle

Cirsium horridulum
Asteraceae (Aster Family)

Flowering Season: All year.

Native Range: Eastern and southern United States, Mexico, and the Bahamas.

Comments: This species is a biennial, flowering in its second year and then setting seed before dying. Although the symmetrical rosettes of grayish-green leaves are attractive and the showy, purplish-pink (rarely white) flowers appeal to an array of butterflies, some consideration must be given to its sharp spines. It is a perfect candidate for a cactus garden or other xeriscapes, because it tolerates dry soils in full sun. In the wild, it can be found in dry to seasonally moist soils. It is a larval host plant of the little metalmark and painted lady butterflies within their Florida range.

A flowering plant can reach 2' tall with a rosette of leaves at the base. Bumblebees and beetles, along with butterflies and hummingbirds, compete for the floral nectar. Purple thistle is not cultivated, so you must collect seeds on your own. Plant seeds each year to ensure you will have flowering plants year round, harvesting seeds before removing spent plants.

Another Florida native species is Nuttall's thistle (*Cirsium nuttallii*), which has equally attractive flowers. It can reach 8–10' in height and is also wickedly spiny.

Blue Mistflower

Conoclinium coelestinum

Asteraceae (Aster Family)

Flowering Season: Spring into winter.

Native Range: Eastern North America from Ontario to Texas, and throughout mainland Florida.

Comments: In cultivation, this attractive Florida native wildflower may reach 12–18" tall and can be covered with small heads of pale blue flowers. It should be regarded as one of the absolute best butterfly-attracting plants in Florida, so put it on your wish list and then seek it out from local nurseries that specialize in Florida native plants.

It performs best in full sun where it can be watered during times of infrequent rainfall—in nature, it grows in short-hydroperiod prairies and other habitats where soils stay relatively moist. It is small enough to be kept in a container situated in a sunny location and will also look right at home in a bed of other herbaceous wildflowers.

If you are into collecting your own seeds, then look for this common wildflower along roadsides that bisect its habitat. In the southern counties, it can be easily confused with the exotic look-alike bluemink (*Ageratum houstonianum*) that is naturalized along roadsides.

Leavenworth's Tickseed

Coreopsis leavenworthii
Asteraceae (Aster Family)

Flowering Season: All year.
Native Range: Florida and Alabama.
Comments: All 11 of Florida's native species of *Coreopsis* were chosen to represent Florida's state wildflower, and Leavenworth's tickseed is the common species throughout much of the state and the only species in the southernmost counties. It is abundant in wet flatwoods and flowers profusely in areas that have burned in recent months, sometimes turning vast fields into a sea of yellow.

Leavenworth's tickseed has made it into the nursery trade, so look for it in nurseries that specialize in Florida native plants. It will readily reseed in moist areas where there is exposed soil, or you can collect your own seeds and sow them wherever you want to add a splash of color. It is an annual, so be certain to collect seeds each year or simply allow it to reseed where it wants.

Small butterflies, especially hairstreaks, visit the tiny disk flowers. This plant grows in wet flatwoods where there is strong sunlight, so this is what you need to duplicate in your home garden. It can easily be grown in a container or a raised bed and would make an attractive border along entryways or around the edge of a sunny patio. There are 14 species found in Florida including the 11 that are native to the state, so you have others to choose from if you can locate them. One species (*Coreopsis nudata*) has pink flowers.

Bat Face

Cuphea llavea

Lythraceae (Loosestrife Family)

Flowering Season: All year.

Native Range: Mexico.

Comments: This subshrub grows to about 24" tall. In addition to having the cutest flowers you can imagine, it flowers freely and continually on new growth throughout the year. The flowers are said to resemble the face of a bat, hence its common name. It is hardy in Zones 10 and 11 but is grown as an annual in cold temperate regions.

Butterflies are attracted to the flowers, as are hummingbirds, but to be more accessible to hummingbirds it should be grown in an elevated bed or even in a hanging basket. It can be planted in full sun or partial shade, but the soil needs to be kept moist or it will wilt and look bedraggled. It is small enough to keep in a container and will be a conversation piece whenever guests sit near it on your patio.

A related species, cigar flower (*Cuphea ignea*), is very similar but with tubular flowers that resemble a lit cigar. Look for both species in local garden centers, or search through the abundance of sources that offer them on the Internet. Yet another species to look for is Mexican heather (*Cuphea hyssopifolia*), which has violet flowers attractive to butterflies. All can be used as border plants or to line a patio.

Tall Elephantsfoot

Elephantopus elatus
Asteraceae (Aster Family)

Flowering Season: Spring through fall.
Native Range: Southeastern United States.
Comments: This is the only species in the genus that ranges south into the Corkscrew Swamp region and the Big Cypress National Preserve. The unmistakable basal leaves are very large, reaching up to 10" long and 3" wide. A tall spike arises from the center of the rosette of leaves and averages about 2–3' tall but may be taller. There are 3 triangular leaflike bracts topping the stems, and these subtend small but interesting pale violet flowers.

Sit and watch wild plants of tall elephantsfoot for awhile, and it won't be long before you see butterflies, the species ranging from large swallowtails to small hairstreaks and skippers. If you have an open, sunny spot in your garden with sandy, moist soil, then tall elephantsfoot will thrive.

There are three other species native to Florida, but this is the most common. Check nurseries that specialize in Florida native plants and keep your fingers crossed. A gray hairstreak is on the flowers in the photo. *Elephantopus* means "elephant foot" and is believed to be an aboriginal name of a species in India.

Coker's Golden Creeper

Ernodea cokeri
Rubiaceae (Madder Family)

Flowering Season: All year.
Native Range: Bahamas and South Florida.
Comments: Despite the existence of herbarium specimens collected in South Florida as early as 1904, this species, previously thought endemic to the Bahamas, was not recognized as a Florida native until 1996. The genus name *Ernodea* is Greek for "offshoot" and refers to the many branches produced by the plant to form compact, leafy mounds. Coker's golden creeper is found in pine rockland habitat in southern Miami-Dade County and on Big Pine Key in the Lower Florida Keys (Monroe County).

The small, tubular flowers open white and then turn pink, creating an attractive arrangement. I have discovered that if it is grown in a hanging basket and placed in the open, ruby-throated hummingbirds are more than happy to sip nectar from the flowers. If grown as a groundcover, it will be too low for hummingbirds.

A closely related native species, *Ernodea littoralis,* can also be used in the same manner; it is more common in the Florida nursery trade. In fact, you will likely not find Coker's golden creeper in nurseries outside of Southeast Florida (yet).

Whichever species you grow, use it as a groundcover or to line entryways, or let it cascade out of a hanging basket for the hummingbirds. Butterflies visit the flowers wherever you use it in your garden.

Green Leopard Plant

Farfugium japonicum
Asteraceae (Aster Family)

Flowering Season: Late summer into fall.
Native Range: Japan.
Comments: In its native Japan, this species grows near wooded stream banks, so you must provide it with shade and reliably moist soil. It will wilt quickly if the soil is not kept moist, and it will not tolerate full midday or afternoon sun. It performs best in a situation where it receives sun for a few hours in the morning and then light shade the rest of the day.

It is an attention grabber in the landscape and is grown as much for its handsome leaves as it is its yellow, daisylike flowers that tower 18" above the foliage. Some gardeners even cut the blooms off so they do not detract from the foliage. The flowers are typical of the aster family with yellow, spreading ray flowers and small yellow disk flowers that butterflies visit.

The leaves are entirely unlike those of other asters; they resemble lily pads. A cultivar named 'Aureomaculatum' has white spots covering the leaf blades and is the source of the name leopard plant. Another cultivar, 'Argenteum,' has white variegation in the leaves, and 'Giganteum' has leaves to 15" across.

Coastal Plain Yellowtops

Flaveria linearis
Asteraceae (Aster Family)

Flowering Season: All year.
Native Range: Florida, Bahamas, West Indies, and Mexico.
Comments: This Florida native perennial is ubiquitous in coastal saltmarshes, flatwoods, wet prairies, and open, disturbed sites such as roadsides that bisect its natural habitat. The flat-topped inflorescence comprises hundreds of small yellow flowers that are very attractive to butterflies.

Coastal Plain yellowtops requires full sun and looks best in beds or mixed in with other herbaceous wildflowers. It prefers moist soil and is easy to grow if given a sunny location. It can be kept in a container if you are diligent about watering. Unlike many native wildflowers, this species is available from nurseries in Central and South Florida that specialize in native plants.

Vast fields of this plant can be seen in open prairies throughout much of Florida, especially within a few months after fire. This cheery wildflower should be a welcome addition to your garden or patio planter.

Blanketflower

Gaillardia pulchella
Asteraceae (Aster Family)

Flowering Season: All year.

Native Range: Maine to Florida and west to South Dakota and Arizona.

Comments: Blanketflower is a staple wildflower in many Florida gardens, so it hardly needs any introduction. Although it is an annual, it will persist in gardens by continuously reseeding, especially in areas with bare soil. Within its range it is also known as firewheel and Indian blanket and is widely available in seed packets at garden centers everywhere.

The seeds can be sprouted in pots or flats or be spread directly in the garden wherever you want a patch of colorful daisies. Butterflies cannot resist the small, yellow disk flowers, and birds such as painted buntings, indigo buntings, and sparrows delight in eating the seeds.

Flower color ranges widely, but the ray flowers ("petals") are typically two-toned with red and yellow. They can also be solid red, solid yellow, or even purplish red. Search the Internet for seeds of various color forms.

The only thing blanketflower requires is full sun all day long and well-drained soil. In the wild it prefers sandy habitats but is quite at home growing along roadsides. It is often planted to beautify Florida roadways.

Beach Verbena

Glandularia maritima
Verbenaceae (Verbena Family)

Synonyms: *Verbena maritima.*
Flowering Season: All year.
Native Range: Endemic to Florida.
Comments: Floridians looking for a native wildflower that attracts butterflies galore need look no further. Plants in this entire family are renowned as butterfly attractors, and beach verbena doesn't disappoint. The bountiful and attractive flowers are produced throughout the year. Its natural habitat in Florida is dunes and coastal pinelands, so it is both drought- and salt-tolerant—two major pluses for coastal residents.

To thrive, beach verbena requires an open, sunny situation with well-drained soil. If this is not available in your garden, then try growing it in a hanging basket so the flowering stems can cascade over the rim of the pot. In order to avoid leaf fungi, avoid regular overhead irrigation and keep the soil on the dry side.

Butterflies constantly visit the flowers, so this is a real champion for butterfly gardening. It is widely available through nurseries that specialize in Florida native plants. Check your local chapter of the Florida Native Plant Society for sources. It is also called coastal mock vervain.

Southeastern Sunflower

Helianthus agrestis
Asteraceae (Aster Family)

Flowering Season: Summer through fall.

Native Range: Eastern United States south through much of the Florida mainland.

Comments: This and other native members of the genus bring bright color to Florida's roadsides and sunny meadows, and they will do the same in your home garden if their simple needs are met. Southeastern sunflower grows naturally in wet marshes and flatwoods, so it will require regular irrigation to help simulate its native habitat.

There are more than a dozen native *Helianthus* in Florida, but southeastern sunflower is one of only a handful to catch the attention of nursery growers who specialize in native wildflowers. Its popularity is no wonder, since its heads of bright yellow flowers not only please gardeners but also attract nectar-seeking butterflies and other insects.

What many people view as a single "daisy" flower is actually a composite of many flowers: an outer ring of ray flowers (whose elongated tubular corollas look like "petals") surrounding many tiny disk flowers (whose tubular corollas are typically short and inconspicuous). The showy ray flowers offer a convenient landing pad for butterflies to access the nectar from the disk flowers.

Other species to consider are narrowleaf sunflower (*Helianthus angustifolius*) and the Florida endemic flatwoods sunflower (*H. carnosus*). Simply add sun and water, and then wait for showtime. 🌿 🦋

Mexican Bell Flower

Hesperaloe campanulata
Agavaceae (Agave Family)

Flowering Season: Sporadically throughout the year, but mostly summer through late fall.

Native Range: Mexico.

Comments: This member of the century plant family is a showstopper when in flower, but it requires very dry, well-drained soil in full sun. If this cannot be provided in the ground, then it should be grown in a container. The stiff leaves are very narrow and reach about 24" long, forming a tight rosette. It will eventually produce offsets and form a cluster of plants. Unlike many members of its family, the leaves of this species are not spiny. The flower spikes can reach 10' tall. Very ornamental, pink-striped bells hang from the branches of the inflorescence, and these are a favorite of hummingbirds wherever the plant grows.

This is a desert plant and is common in the Chihuahuan Desert in Central Mexico. Try a standard cactus soil mix with additional perlite or coarse gravel, and keep it in full sun as if it were in a desert. In Mexico, it grows at altitudes up to 1,800 feet, which should make it relatively cold tolerant. It is cultivated at Fairchild Tropical Botanic Garden in a raised bed of gravelly soil, where it was photographed for this guide. You can buy seeds and plants on the Internet.

Camphorweed

Heterotheca subaxillaris
Asteraceae (Aster Family)

Flowering Season: All year.
Native Range: Delaware to Texas, Mexico, and throughout mainland Florida.
Comments: Camphorweed is a common constituent of Florida's pine flatwoods, sandhills, and dunes, where it can be found flowering throughout the year. When crushed, the leaves smell very strongly of camphor, which is a key to telling it apart from other yellow-flowered composites that share its habitat.

Unless you can find camphorweed cultivated in a local nursery that specializes in native plants, you will likely be on your own collecting seeds. This should not pose much of a dilemma, because it is frequent throughout much of Florida in many unprotected areas like roadsides and other disturbed sites, or on private landholdings where you may be able to collect seeds with the owner's permission. It is worth seeking out, because the yellow, daisylike flowers are attractive to small butterflies, and it is trouble-free once it is established in a sunny location. It will likely even spread to other areas in your garden wherever there is exposed soil.

A similar native species that is also a worthwhile addition to your wildflower garden is silkgrass (*Pityopsis graminifolia*), which has silky, grasslike leaves and slightly smaller flowers. The two species are sometimes found growing together in the same habitat.

Coral Senecio

Kleinia fulgens

Asteraceae (Aster Family)

Synonyms: *Senecio fulgens.*

Flowering Season: Spring to early winter.

Native Range: Africa.

Comments: Meet one of the very few succulents in the aster family. Coral senecio is right at home alongside cactus in a sunny bed of fast-draining, sandy or gravelly soil. The gray-green leaves contrast nicely with the rounded clusters of burnt orange flowers, and butterflies simply cannot resist their offerings of nectar and of alkaloids that they use to make pheromones for mating. Give this species a long resting period through winter without any supplemental irrigation, and it will burst into full glory come the first spring rains. In proper conditions, it will spread from rhizomes and form an extensive colony.

If you cannot meet its horticultural needs in the ground, then prepare a raised bed of cactus soil that has been further amended with coarse sand. Do not use playground sand or builder's sand because it will not drain well. Excellent coarse sand can be found at businesses that rent sandblasters. For even better drainage, add perlite or some coarse gravel. Your best source for coarse gravel will be at your local aquarium store if you only need small amounts.

Coral senecio is very easy to root by simply cutting stems into sections and sticking them in a pot of soil. It is not a species that you will find in mainstream nurseries, but the South Florida Cactus and Succulent Society sponsors sales at Fairchild Tropical Botanic Garden. You can also look for plants from Internet sources.

Rockland Lantana

Lantana depressa var. *depressa*
Verbenaceae (Verbena Family)

Flowering Season: All year.

Native Range: Southern Miami-Dade County, Florida.

Comments: This Florida endemic has been a parent of many hybrids in the nursery trade, often creating sterile plants with a mounding or trailing growth habit. These include such notable hybrids as 'Gold Mound,' 'New Gold,' 'Lola,' and 'Lemon Swirl.'

As a species, rockland lantana forms low mounds of stems in pine rockland habitat in and to the east of Everglades National Park. In many areas its gene pool has been compromised through natural hybridization with the invasive exotic *Lantana camara*. Good luck finding pure, unadulterated plants of *Lantana depressa* for your South Florida butterfly garden. Look for it in Miami-Dade County nurseries that specialize in Florida native plants, because many other nurseries sell hybrids that are mislabeled as this endemic species, especially 'Cream Carpet,' which bears pale yellow or nearly white flowers with a dark yellow center.

Rockland lantana requires full sun and dry, well-drained soil and would make a nice container plant for a sunny deck or patio. Its flowers are always rich butter yellow, turning uniformly orange yellow with age. 🦋

Lantana 'Lola'

Lantana 'Lola'

Verbenaceae (Verbena Family)

Flowering Season: All year.

Native Range: Of cultivated hybrid origin.

Comments: This relatively new introduction belongs to the Lantana Callowiana Group of hybrids that includes some fascinating sterile, mounding plants for sunny gardens. The bushy wild forms of *Lantana camara* in Florida are not recommended due to their weediness in natural areas, their very poisonous fruits, and their propensity to hybridize with an endangered South Florida endemic species called *Lantana depressa*.

Dr. Roger Sanders recently determined that the wild, escaped plants in Florida are not *Lantana camara* but rather an undescribed species that he named *Lantana × strigocamara*. Regardless, as a species it is listed as one of Florida's most invasive plants. If you have children, dogs, or livestock, you and they will be better off without it.

The good news is that 'Lola' and other Callowiana Group hybrids in the nursery trade are sterile, so they do not pose a threat to natural areas or your children. The leaves, however, are still toxic to dogs. Other sterile hybrids include 'Pink Caprice', 'Dwarf Pinkie', 'New Gold', 'Gold Mound', 'Patriot', 'Weeping White', 'Weeping Lavender', 'Texas Flame', and 'Lemon Swirl'. These are all superior plants for attracting butterflies. Give them full sun and dry, well-drained soil. 🦋 ☠️

Blazing Star

Liatris spp.

Asteraceae (Aster Family)

Flowering Season: Spring and summer.

Native Range: There are 17 species native to Florida; 4 of them are endemic to the state.

Comments: All members of this genus are superb attractors of butterflies, so no matter which species you grow, the butterflies will be all over the flowers. They are unusual in that the flowers open from the top of the spike downward. When you find patches of blazing stars in the wild, you will find plenty of butterflies ranging from large swallowtails to small skippers and everything in between. Hummingbirds visit the flowers of the taller species.

If you do not have a sunny place to grow blazing stars in the ground, then simply keep them in containers. They are perennials but will die back in winter and sprout again in spring. Most species have pinkish-purple flowers with occasional white morphs, but *Liatris elegans* of northern Florida produces white flowers and is stunning to see in the wild.

Several species of Florida's native blazing stars grow in scrub and sandhill habitat and will require well-drained sandy soil. The species pictured is *Liatris spicata*, which is locally common throughout Florida. It and many other species are available in seed packets, so check your local garden centers. Seeds should be sown either directly in the ground where there is exposed soil in a sunny location or planted in flats and transplanted later. Another common name is gayfeather.

Cardinal Flower

Lobelia cardinalis
Campanulaceae
(Bellflower Family)

Flowering Season: Summer through fall.

Native Range: Southeastern Canada south through the eastern and southwestern United States and south through Mexico into Central America.

Comments: Words cannot describe the vision of paddling a canoe along one of the scenic rivers in northern Florida and seeing cardinal flower blooming along the banks. The spikes can reach up to 4' tall, and the lipstick-red flowers stand like brilliant sentinels for every passing butterfly, hummingbird, and canoeist.

Cardinal flower is widely cultivated and was introduced into England in the early 1600s, where it was named because of the resemblance of the flower color to the ecclesiastical garments worn by Roman Catholic cardinals.

It must be grown in reliably moist, acidic soil, so if you do not live on a scenic river, plant it in a water garden or in a pot sitting in water. Hummingbirds are almost the exclusive pollinator, which alone should inspire you to grow this beautiful wildflower. Collect seeds from your plants every year, because it is short-lived, and some plants die after flowering.

Besides nurseries that specialize in Florida native plants, Internet sources have it widely available, or if you live within its natural range in Florida, you can look for it in flooded roadside ditches to collect seeds. In addition to hummingbirds, swallowtails and other butterflies visit the flowers for nectar.

Sunshine Mimosa

Mimosa strigillosa

Fabaceae (Pea Family)

Flowering Season: Spring to fall.

Native Range: Florida, Texas, and Mexico.

Comments: If you happen to be in the market for a very pretty, fast-growing groundcover that will take light foot traffic and is a Florida native, then please allow me to introduce you to sunshine mimosa. It ranges south into Collier County and has gained popularity in the nursery trade in recent years as an alternative to turf grass in small areas. The flowers form oblong clusters of bright pink stamens, and the leaves are divided into many small, narrowly linear leaflets that close when touched. The stems stay snug to the ground and branch profusely in a crisscrossing pattern.

In addition to its usefulness as an attractive groundcover, it is also a larval host plant of the little yellow, a tiny butterfly often seen flying low to the ground. A variety of butterflies visit the flowers for nectar, as well.

Sunshine mimosa can be planted wherever you have bare soil, but it does require full sun to succeed in cultivation. You can even try it in a wide hanging basket. It is so popular you can sometimes find it in mainstream garden centers. It is closely related to the Florida native sensitive brier (*Mimosa quadrivalvis*), but that plant has smaller clusters of flowers, and its stems are wickedly spiny.

Parsley

Petroselinum crispum

Apiaceae (Celery Family)

Flowering Season: Fall into spring.

Native Range: Middle East.

Comments: We all know that Parsley is useful as a culinary herb and decorative garnish, but a lesser-known fact is that it is a larval host plant of the black swallowtail butterfly (larva pictured). Other members of this family, especially dill and fennel, are used as larval food by this well-known butterfly. Because parsley is a kitchen herb it is hardly thought of as a landscape plant, but it is quite decorative when used as a border plant to line short walkways. It can be useful in many other garden situations and can even add its delicate green foliage to a mix of flowers in a sunny window box. The small white flowers may even attract beneficial insects like lady beetles to your garden.

The flowers are in umbels, but most gardeners harvest the leaves for kitchen use long before it matures enough to flower. It is not a long-lived plant and is typically grown as an annual or biennial. Black swallowtail butterflies in southern Florida lay their eggs on the leaves in winter and spring, so plan on growing enough of it to share with them. Seeds are available in packets, and young plants can be purchased in the vegetable and herb section of garden centers throughout Florida.

Creeping Charlie

Phyla nodiflora
Verbenaceae (Verbena Family)

Synonyms: *Lippia nodiflora.*
Flowering Season: All year.
Native Range: United States through tropical America and the West Indies.
Comments: The oft-quoted saying about one man's weed being another man's wildflower certainly applies here, because creeping Charlie is one reason many people use products that kill weeds in their manicured lawns. They would be flabbergasted to learn that specialty nurseries now grow it for sale. Goodness gracious, how times have changed. Actually, creeping Charlie would make a better groundcover than lawn grass, and you could display your retired lawnmower outside as yard art.

If some enterprising nursery would grow creeping Charlie in flats, it could be laid out just as you would sod. The tiny flowers are formed in a concentric circle around the head; they are white when they first open but later turn pinkish purple. Creeping Charlie is an important larval host plant in urban areas for the white peacock butterfly; it also provides nectar for small butterflies like crescents, blues, hairstreaks, and metalmarks. Its other common names include capeweed, matchsticks, fogfruit, and frogfruit.

Propagation is by seed, but it's easy to simply dig up clumps of plants along roadsides, abandoned fields, or in your neighbor's lawn. For a real conversation piece, try growing it in a hanging basket so it can cascade over the edge. Your houseguests will think you've gone over the edge, too.

Obedient Plant

Physostegia virginiana
Lamiaceae (Mint Family)

Flowering Season: Late spring to early fall.
Native Range: Northeastern United States.
Comments: The common name comes from the flowers' odd characteristic of staying in whatever position you bend them, so here's a plant that entertains children and also attracts hummingbirds and an array of butterflies. There are three Florida native *Physostegia* species, but they are not seen in cultivation as often as this attractive plant. Lovers of Florida native plants should look for eastern false dragonhead (*Physostegia purpurea*) in nurseries that specialize in native wildflowers. It has narrower pinkish to purple flowers.

Obedient plant has escaped from cultivation from north-central Florida to the Panhandle, but it is not an invasive pest. The showy flowers are massed along upright stems to 3' tall and look exceptional when intermixed with other wildflowers or planted in masses. Skipper butterflies are especially adept at crawling down the floral tube to gain access to the nectar.

It is hardy throughout Florida and can be found in local nurseries, or you can readily purchase seeds and plants from Internet sources. Obedient plant benefits from regular watering and requires full sun. There is a white-flowered form in cultivation called 'Miss Manners.'

Scarlet Plumbago
Plumbago indica
Plumbaginaceae
(Plumbago Family)

Flowering Season: Late fall to spring.

Native Range: Southeast Asia.

Comments: Scarlet plumbago is so outstandingly excellent for garden decoration that the British Royal Horticultural Society gave it their Award of Garden Merit. It is hardly known in cultivation in the United States but can be easily acquired from Internet sources.

It is a reliable bloomer and adds color to Florida gardens in winter. It has been described as "vigorous" and "rambunctious," so keep your garden shears handy. Pruning it in late spring will encourage new growth that will produce an abundance of brilliant blooms on 12" spikes beginning in fall and lasting into spring. The festive flowers attract hummingbirds and butterflies, and the leaves serve as larval food for the Cassius blue butterfly.

It looks quite ornamental when planted with the white-flowered Florida native doctorbush (*Plumbago zeylanica*) because they both have a similar growth habit and the flowers contrast nicely.

Gardeners can take advantage of its rangy growth habit by allowing it to scramble up into shrubs with its flowering stems sticking out into the sun.

Doctorbush

Plumbago zeylanica
Plumbaginaceae
(Plumbago Family)

Synonyms: *Plumbago scandens.*
Flowering Season: All year.
Native Range: Coastal southern United States, West Indies, and tropical America to tropical Asia.
Comments: This obscure Florida native is found along sandy coastlines, where it grows along dunes. The flowers attract small butterflies, and the leaves are a favorite larval food of the Cassius blue butterfly. Although the blue-anthered flowers are attractive, this wildflower is seldom appreciated because of its sprawling growth habit and fruits that cling to hair and clothing. If you have a sunny area with dry, well-drained soil, it is a worthy garden subject. Be advised that if you have pets with long hair, the fruits can become a terrible nuisance if they run through patches of this plant.

Although a few nurseries that specialize in Florida native plants offer this species, it is easily obtained by collecting seeds from weedy coastal areas, especially along back dunes of beaches. Simply run festively through thickets of this plant, and then pick the seeds from your socks, or let your dog be the seed collector. Bring your hair clippers, or better yet, borrow your neighbor's dog. The common name relates to its medicinal uses throughout its wide native range.

Pickerelweed

Pontederia cordata
Pontederiaceae
(Pickerelweed Family)

Flowering Season: All year.

Native Range: Virginia to Missouri and Texas, south through Florida and tropical America.

Comments: Why pickerelweed doesn't receive standing ovations among Florida gardeners as a butterfly attractor is a great mystery. The spikes of showy blue (rarely white) flowers are butterfly magnets, and it is exceptionally easy to grow if you have a water garden; you can also simply plant it in a large pot and sit the pot in a tray of water so the soil remains wet. Preformed plastic tubs designed for water gardens can be used to hold water with submerged pots, or you can try filling it with soil and then running your garden hose in it until it overflows. You now have a freshwater wetland.

Pickerelweed spreads from rhizomes and will quickly outgrow pots but will happily spread across the surface of water. Nurseries that specialize in water gardens will have this plant for sale, but digging up a clump from roadside ditches or along canal banks saves at least those few plants from death by herbicide. It is very common throughout mainland Florida but does not receive the horticultural attention it deserves as a butterfly attractor.

Thickleaf Wild Petunia

Ruellia succulenta

Acanthaceae (Acanthus Family)

Flowering Season: All year.

Native Range: Endemic to southern mainland Florida.

Comments: The common buckeye butterfly uses native *Ruellia* species as larval food statewide, but in southern Florida the leaves of this endemic species offer larval food for the beautiful malachite butterfly as well. This rare, malachite-green butterfly also uses the non-native green shrimp plant (*Ruellia blechum*), a pernicious weed in much of Florida.

Thickleaf wild petunia forms low mounds of leaves and is trouble-free if it receives full sun and kept free of competition. It sometimes has purple leaves, and there are pink-flowered forms, too. It looks right at home in a wildflower garden, or you can grow it in a container in a sunny location. Skippers crawl down the throats of the flowers, and larger butterflies will probe the flowers with their tongues. The similar Carolina wild petunia (*Ruellia caroliniensis*) occurs from lower Central Florida northward, so it may be more available within its native range.

The commonly cultivated bluebells (*Ruellia brittoniana; R. simplex*) grows to about 3' tall with violet or pink blossoms and is seriously invasive. It should be removed and discarded despite its popularity and temptingly pretty flowers. A non-native species worth growing is the quaint and well-behaved *Ruellia squarrosa.*

Lizard's Tail

Saururus cernuus
Sauraceae (Lizard's Tail Family)

Flowering Season: Spring into fall.

Native Range: Quebec and Ontario across the eastern half of the United States.

Comments: Lizard's tail inhabits Florida's freshwater wetlands, so you will need to create this habitat in your garden to successfully grow this species. This can be easily accomplished by using preformed plastic ponds, a discarded bathtub with the drain hole sealed (if ordinances allow "yard art" in your city), or by using a large plastic pot with the drain holes sealed.

This plant's botanical name translates to "nodding lizard's tail," which perfectly describes the down-swept spikes of flowers that superficially resemble the tail of a lizard. Swallowtails and fritillaries are especially fond of the flowers. It is a handsome plant that will reward you with sweetly perfumed flowers throughout the year. It spreads by underground rhizomes and will claim an area for itself fairly quickly.

Lizard's tail is very cold tolerant and requires no winter protection anywhere in Florida. It is not readily available, but check nurseries that specialize in water garden plants. Also check your local roadside ditches. Its only requirement is permanently wet soil, but it may need mild doses of chelated iron if it becomes iron deficient.

Bolivian Sunset

Seemannia sylvatica

Gesneriaceae (Gesnariad Family)

Synonyms: *Gloxinia sylvatica.*
Flowering Season: Late summer into winter.
Native Range: Bolivia.
Comments: Gardeners, meet one of the cutest plants you will ever see. Bolivian sunset hails from high elevations in the Andes Mountains and is hardy to Zone 8, so its worst enemy in tropical Florida is summertime. Growing it in a hanging basket and moving it to a cooler area in the heat of summer may remedy this problem. A hanging basket also elevates the plant so hummingbirds will find it. As temperatures cool in fall, it explodes into color, showing off its handsome, eye-popping blossoms, until winter cold forces it into dormancy. It produces scaly rhizomes that can be overwintered indoors if kept cool and dry.

Hummingbirds frequent the flowers even when there are other choice blossoms nearby. Butterflies are not as frequent, but they do visit the flowers for nectar occasionally. It grows to about 18" tall and prefers rich soil with regular watering in the dry season, but not when it is dormant.

Look for Bolivian sunset in nurseries that specialize in ornamental plants. There are many other species to choose from, but this species is the best performer in tropical Florida.

Goldenrod

Solidago spp.

Asteraceae (Aster Family)

Flowering Season: Summer and fall, but some species flower all year.

Native Range: Throughout the Americas and West Indies.

Comments: In Florida there are 20 native goldenrod species from which to choose. Some are freshwater wetland species, some grow in saltmarshes and coastal dunes, and others inhabit pinelands and prairies. All produce an abundance of small, butterfly-attracting yellow flowers on erect, often branching stems. There is a common belief that goldenrod flowers cause allergies, but this is entirely false; they do not have airborne pollen. The real culprit is common ragweed (*Ambrosia artemisiifolia*), which flowers at the same time.

The species photographed is seaside goldenrod (*Solidago sempervirens*), a robust species found throughout much of Florida in a wide range of coastal and inland habitats. The large, 8–16" leaves in a basal rosette makes it easy to recognize. Another tall species that does well in cultivation, spreading from rhizomes to form patches of flowering stems, is Leavenworth's goldenrod (*Solidago leavenworthii*). Whichever species you choose, give it a sunny location, but also do some research to determine its preferred natural habitat, so you will know if it requires wet soil or not. Goldenrods all grow easily from seed.

Lopsided Indiangrass

Sorghastrum secundum
Poaceae (Grass Family)

Flowering Season: Summer through fall.

Native Range: South Carolina to Kansas, south to Louisiana, and throughout Florida.

Comments: Grasses offer a distinct look that other plants cannot provide in gardens. Of the many Florida native grasses to choose from in decorating your landscape, lopsided Indiangrass should contend for top honors. It is gracefully attractive in flower, plus it is a preferred larval host plant for a group of butterflies known as grass skippers. These butterflies all use grasses as host plants; those that use lopsided Indiangrass include the swarthy skipper, arogos skipper, dusted skipper, Eufala skipper, and twin-spot skipper.

Native grasses tend to be underutilized in landscape designs; this is due in part to unavailability. But lopsided Indiangrass is occasionally grown by nurseries in Florida that specialize in Florida native plants, so check your local chapter of the Florida Native Plant Society for sources. If you know locations of unprotected sandy pineland parcels, those would be good places to look for this grass and collect your own seeds. Just look for a grass with all of the flowers on one side of the stem.

Stokes' Aster

Stokesia laevis

Asteraceae (Aster Family)

Flowering Season: Summer.

Native Range: Southeastern United States to the central and western Florida Panhandle.

Comments: Years ago, while traveling through the Apalachicola National Forest in the Florida Panhandle, my wife and I came upon a roadside patch of this spectacular native wildflower. Later the following year, we encountered plants for sale at a Florida Native Plant Society state conference and bought one. Today it still resides in a pot sitting half-submerged on the edge of our natural swimming pool, where each year it delights us and the butterflies with its stunning flowers.

It grows in wet flatwoods, bogs, savannas, and along roadsides that bisect its habitat, so it definitely prefers wet soil. It can thrive in a pot but will need regular watering, or you can try placing the pot in a dish of water to ensure that the soil remains moist.

The flower heads are about 2½" wide and attract a wide array of butterflies to the small central disk flowers. Tropical Florida gardeners will likely have to rely on Internet sources to buy it. You can even find some cultivars like 'Blue Danube,' 'Purple Parasols,' 'Honeysong Purple,' and 'Peachie's Pick.' All are worthy of a moist sunny spot in any Florida garden. 🌶️ 🦋

Elliott's Aster

Symphyotrichum elliottii
Asteraceae (Aster Family)

Synonyms: *Aster elliottii.*
Flowering Season: Summer through fall.
Native Range: Southeastern United States.
Comments: Elliott's aster is common throughout much of Florida and sometimes forms spreading colonies in roadside ditches, where it may reach 6' tall or more. When it blooms, from late summer through the cooler fall months, you cannot possibly miss it.

It requires reliably wet soil and may spread well out of bounds if given proper conditions. If you find a wild patch of Elliott's aster in flower, you will find plenty of butterflies, so having it spread out of bounds might not be so bad after all. Check the Florida Wildflower Growers Cooperative for seed packets (see the Resources section) or collect your own seeds from roadside plants growing outside of protected areas. This is the perfect plant for the edge of a lake (if you're lucky enough to have lakefront property), or you can try planting it in an area of your garden where you can run a sprinkler when necessary.

This species is quite similar to Simmonds' aster (*Symphyotrichum simmondsii*), and it can be substituted if you happen upon it in a nursery or a roadside ditch. 🦋

Mexican Tarragon

Tagetes lucida
Asteraceae (Aster Family)

Flowering Season: Midsummer to early winter.

Native Range: Mexico and Guatemala.

Comments: The leaves of this herb have a tarragon-like odor, with hints of anise, and can be used as a substitute for French tarragon in the kitchen. So go ahead and plant it in your herb garden; the butterflies won't mind. It also has a long history of medicinal uses, and in its native lands it was smoked for divinatory purposes during religious ceremonies. Toss some flowers and leaves in your salad or soup for a distinctive taste treat, or use it to prepare béarnaise sauce, rémoulade, and tartar sauce.

In butterfly gardens, it is particularly useful for attracting blues, crescents, hairstreaks, and metalmarks, small butterflies that frequent urban gardens in Florida. Mexican tarragon averages 14–24" tall and may die back in winter, resprouting in spring. The flowers are about ½" wide and literally cover the plant.

Your best sources for locating Mexican tarragon seed will likely be found on the Internet, although some Florida specialty nurseries might have plants. It will perform nicely as a border plant along the south side of a foundation, in a raised bed, or kept in a container on a sunny patio. 🦋

Fiesta Del Sol

Tithonia rotundifolia
Asteraceae (Aster Family)

Flowering Season: Spring into winter.

Native Range: Mexico through Central America.

Comments: This vigorous, bushy annual reaches 3–6' tall, depending on how well it is grown, and it flowers profusely if given full sun and a chance to dry out in between waterings. There are some catchy names in the nursery trade designed to further tempt gardeners into buying plants or seeds, but not much added temptation is necessary once you have seen the vivacious flowers of bright orange or reddish orange. The common name translates to "festival of the sun."

The nectar feast offered by the yellowish-orange disk flowers will have butterflies coming and going all day long. The blossoms measure 2½–3" across and stand out in the garden like lightbulbs. You should be able to find seed packets at your local garden center. Sow the seeds directly into the garden, or start them in flats and transplant the seedlings wherever you want outlandish color. Hummingbirds are also attracted to the blooms. It makes an excellent cut flower, so whatever the seeds cost, you will certainly get your money's worth. 🌺 🦋

Society Garlic

Tulbaghia violacea
Amaryllidaceae
(Amaryllis Family)

Flowering Season: Midsummer into early winter.
Native Range: South Africa.
Comments: Gardeners from around the world are familiar with society garlic because it is an easy-to-grow, vigorous, clumping perennial that sports fragrant, lilac-pink flowers on erect stems through much of the summer and fall. The leaves smell like garlic and can be used in soups and stews.

Gardeners in North America know that it attracts hummingbirds and butterflies, and in England it is so wildly popular that it received the Royal Horticultural Society's Award of Garden Merit. It has recently been found to have anticancer properties and can also be used as an antithrombotic, to ward off blood clots. Use that trivia in your garden tours.

In the garden it will not take up much space, but it does require full sun and soil that drains well. It will serve gardeners well for general decorative uses, wherever splashes of color are desired. Plant it around your mailbox post, in a rock garden, or in beds on either side of an entryway. Its uses are endless.

Department store garden centers often sell society garlic, and there is a handsome form with white-edged leaves called 'Silver Lace.' Society garlic is cold hardy into Zone 7.

White Crownbeard

Verbesina virginica
Asteraceae (Aster Family)

Flowering Season: Late spring through fall.

Native Range: Pennsylvania south through mainland Florida.

Comments: If you are a Florida native-plant aficionado, then you will want this free-blooming plant in your garden. And ditto if you are a butterfly lover, because white crownbeard blossoms are a popular nectar source for a wide variety of butterflies, including the rare Atala butterfly of southeastern Florida.

White crownbeard can reach 6' tall and has erect, distinctly winged stems that are topped with showy clusters of white, daisylike flowers. The central disk flowers are what draw in the butterflies for a nectar banquet.

In the wild it inhabits sandy pinelands, the edges of tropical hardwood hammocks, and coastal strand. This means it will tolerate full sun as well as light, transitional shade in garden settings. It dies down to the ground in winter and re-sprouts with vigor in late spring or early summer, growing rapidly until it reaches flowering size. Plant it in your garden where it can overtop smaller plants. It is available from nurseries that specialize in native plants and is propagated by seed. Another common name is frostweed.

Coontie

Zamia pumila
Zamiaceae (Zamia Family)

Synonyms: *Zamia integrifolia*, illegitimate; *Zamia floridana*.
Flowering Season: All year.
Native Range: Florida and Georgia through the Bahamas and West Indies.
Comments: Cycads are a primitive group of plants whose heyday was during the Jurassic period, and this is the only cycad that is native to Florida. The underground stem is very poisonous if eaten raw but was a staple starch for Native Americans and early settlers once the toxins were washed away with water.

Coontie can tolerate full sun or the shade of trees and can be planted in groups, used to line walkways, or scattered around in the garden. Some forms reach 12" tall, while others may grow four times that height.

Scale insects can be controlled with horticultural oil, by cutting off infested leaves, and by encouraging biocontrols like lady beetles, lacewings, and parasitic wasps. It also helps to eradicate fire ants and avoid using insecticides.

In southeastern Florida, coontie is the only native larval host plant of the rare Atala butterfly. The larvae are red with a row of yellow spots down the sides; the adults have midnight blue wings with iridescent blue spots and orange abdomens. Once you have plenty of coontie planted, you may be able to obtain larvae through the Miami Blue Chapter of the North American Butterfly Association in Miami. 🌴 🐛 ☠

3

The Hummingbirds

Neither the hummingbird nor the flower wonders how beautiful it is.

Emily Lewis, Pleasure Notes blog

Resting between visits to flowers, a rufous hummingbird perches on a twig in the author's garden.

They say hummingbirds represent love, and there is certainly no other bird more lovely and mesmerizing to see in your garden. They demand your full attention by performing a fascinating aerobatic show as they zip from one flower to the next, hovering in one spot, flying backward, or even twirling upside down to access flower nectar. You will find that hummingbirds often visit the same flowers as butterflies, but this is not always the case. Some plants produce flowers specialized to attract hummingbirds as pollinators while preventing access to the nectar by butterflies. This should remind you that variety is the key to creating a successful hummingbird and butterfly garden.

Interestingly, hummingbirds often visit the blossoms of plants that occur naturally in parts of the world where there are no hummingbirds at all. For instance, ruby-throated hummingbirds in Florida are very fond of the nectar produced by flowers of the Chinese hat plant (*Holmskioldia sanguinea*) native to tropical Asia, yet hummingbirds are strictly a New World species and occur only in North America, Central America, South America, the Bahamas, and the Caribbean.

While it is true that hummingbirds are particularly attracted to the color red, some of their favored flowers are purple, violet, yellow, orange, white, or even green. There is also a misconception that hummingbirds only visit tubular flowers, when in reality they visit flowers that run the gamut of shapes and sizes. Also, they are not always seeking nectar. Hummingbirds eat high-energy pollen and soft-bodied insects such as mealybugs and soft scale insects. They are adept at removing trapped insects from spider webs, and they even use spider webs to construct and secure their Lilliputian nests. Besides being crafty little birds, they can attain speeds up to 70 miles per hour. Many people are flabbergasted to discover that hummingbirds may travel thousands of miles during migration. Rufous hummingbirds migrate from the west coast of the United States to the southern tip of Florida each year. Hummingbirds sometimes cross the Gulf of Mexico and the Straits of Florida but even at 50 miles per hour the Bahamas are only an hour away and Cuba is barely a two-hour flight from Florida. At one time people believed hummingbirds migrated by perching on the backs of vultures and other large migratory birds.

Make it a practice to closely observe hummingbirds in your garden. Even though 99 percent of them in Florida will be ruby-throated hummingbirds, stray migrants from other parts of the United States or from the tropics may pleasantly surprise you someday. In the southernmost Florida counties, pay special attention to any hummingbird you see during the summer months, June through August, when hummingbirds are mostly absent. If you see one then, it will likely be a tropical migrant. And remember: if you do find one of the real rarities visiting your home garden and you alert the birding world by posting the sighting on a local or national Rare Bird Hotline, expect some visitors. In fact, expect a *lot* of visitors, and consider charging admission!

Detailed descriptions of the various hummingbirds known to occur in Florida can be found in many bird guides. See the Resources section in the back of this book for recommendations. The following list provides information on noteworthy hummingbirds that have been observed in Florida, listed according to the numbers of documented sightings.

Ruby-Throated Hummingbird (*Archilochus colubris*)

This is the most common and well-known hummingbird in Florida. It breeds throughout the eastern half of the United States and lower Canada from Nova Scotia east into Alberta, south along the east coast to lower Central Florida, and south from eastern North Dakota to eastern Texas. It has been documented nesting in the Big Cypress National Preserve in recent years. It winters in southern Florida and from Oklahoma and Texas through eastern Mexico. Occasionally some birds may overwinter in Cuba and the Bahamas.

In the southern counties of Florida, ruby-throats usually show up in late August or early September and migrate out of the area in April and May. Therefore, it is important to cultivate plants in this region that bloom in fall, winter, and spring.

Rufous Hummingbird (*Selasphorus rufus*)

The breeding range of the rufous hummingbird extends from southeastern Alaska, southern Yukon, British Columbia, and southwestern Alberta south to northwestern California and southern Idaho. Its wintering range is in Mexico south to Guerrero and northern Oaxaca, but sightings occur in virtually every state in the United States east of the Rocky Mountains. They are uncommon but regular winter visitors in tropical Florida each year, typically arriving by early September and remaining through winter. A mature male rufous hummingbird is the only reddish-brown species that can be seen in Florida, so they are unmistakable even for amateur birders. Immature males and adult females are essentially metallic bronze green above with a white belly and rufous sides.

Buff-Bellied Hummingbird (*Amazilia yucatanensis*)

This hummingbird breeds from the lower Rio Grande Valley through eastern Mexico, the Yucatan Peninsula, Chiapas, and Belize. It appears throughout the year in Texas. There are documented sightings in Louisiana and a dubious record from Massachusetts. There are confirmed records in Ft. Lauderdale from 1989 to 1993, Palm Beach County in 2000, and at Fairchild Tropical Botanic Garden in Coral Gables in 2005 and 2006. Extralimital sightings of birds such as the buff-bellied hummingbird may reflect the growing popularity of birding throughout the United States. With more and more birders, new sightings would seem almost inevitable, but this still should be regarded as a rare hummingbird in Florida.

Black-Chinned Hummingbird (*Archilochus alexandri*)

Black-chinned hummingbirds winter almost solely in Mexico, but there are sightings of them east of the Mississippi River in Arkansas, Louisiana, Mississippi, Alabama, North Carolina, Georgia, and Florida. It is certainly not a bird to be expected in tropical Florida but is listed here solely on documented occurrences in southern Florida in recent years. At certain angles the throat feathers of ruby-throated hummingbirds can look black, so keep that in mind if you think you see a black-chinned hummingbird.

Bahama Woodstar (*Calliphlox evelynae*)

The breeding range of the Bahama woodstar encompasses the Bahama Archipelago, where it is mostly sedentary. It rarely crosses the Straits of Florida and was first documented in Florida (as a dead bird stuck in a window screen) in Miami-Dade County in January 1961. The first living bird observed in Florida was in Palm Beach County between August and October 1971. The next report came from Homestead (Miami-Dade County) in April and May 1974. The first pair (male and female) showed up in July 1981 at the Mary Krome Bird Refuge in Homestead. Whether or not the Bahama woodstar is more regular in tropical Florida than sightings indicate is unknown, but considering the popularity of birdwatching, it is probably not very regular. It is the most common hummingbird in the Bahamas, especially around Nassau.

Cuban Emerald (*Chlorostilbon ricordii*)

The Cuban emerald is a resident of Cuba and the northern Bahamas, where it mostly occurs on Abaco, Andros, and Grand Bahama. Like the Bahama woodstar, it very rarely strays into Florida. It was recorded in Florida as early as 1943, then again in 1953 south of Cocoa (Brevard County). More recently there have been confirmed sightings from Naranja (Miami-Dade County) in January 1961, Cocoa Beach (Brevard County) in 1963, and at Hypoluxo (Palm Beach County) in 1977.

Other hummingbirds that have been documented in Florida include Allen's, calliope, broad-tailed, broad-billed, Anna's, and white-eared. All would be exceptionally rare to see in tropical Florida.

Where to See Wild Hummingbirds

If you are not familiar with hummingbirds and would like to get acquainted with them, public botanical gardens are a good place to look during the season when hummingbirds are in your area. You can, of course, find wild hummingbirds in state and national parks and preserves, but they are generally not as easy to locate as they are in botanical gardens. Hummingbird and butterfly gardens have been established at many nature centers, local parks, and even around schools, libraries, and churches.

Check the list of parks and preserves in the back of this guide for

A female ruby-throated hummingbird in the author's garden.

suggested locations to visit. Also consider joining guided field trips sponsored by various Audubon Society chapters. There are even professional guides that can be retained for a fee. Many national, state, and county parks also offer guided bird walks.

Once you have acquainted yourself with the rapid-fire twittering made by hummingbirds, you will be better prepared to notice them around your home garden. They are usually heard long before they are seen. Once they find a reliable source of nectar, they will be back in years to follow. If your garden is especially enticing, then perhaps they'll bring their friends.

Cautionary Advice about Artificial Feeders

Artificial sugar-water hummingbird feeders are not always your best choice. They require frequent maintenance, and the sugar-water solution can be harmful to hummingbirds if not prepared properly. If you do decide to use artificial feeders around your garden, there are some precautions you should take in order to make them as safe as possible for the birds:

- Always boil homemade solutions prior to use, and never mix the solution stronger than 4 parts water to 1 part white granulated sugar. Never use honey. Bring the water to a full boil to kill bacteria, add and dissolve the sugar, cool the solution quickly, and refrigerate unused portions. Stronger solutions can cause liver damage to hummingbirds.
- Do not purchase commercial sugar-water solutions that have red dye added. The dye is not needed to attract hummingbirds and may be harmful to them.
- Change the solution at least once weekly, and scrub the feeder with a mixture of hot water and vinegar before refilling. Never use soap.
- *Never* use insecticides to control bees, wasps, or ants on or anywhere near the feeder.

Once you have an established garden, you will most likely find it more exciting and aesthetically pleasing to watch hummingbirds dart among flowers than visit an artificial feeder. In my opinion your money and efforts are better spent on plants. Consider putting out artificial hummingbird feeders only in the event of a severe winter freeze that damages the flowers of their favorite plants, and then hang the feeders on or near the plants they were accustomed to visiting. In the absence of flowers, artificial feeders do offer hummingbirds a dependable food source, but use them with due caution.

4

The Butterflies

Tropical Florida boasts a number of butterflies that cannot be found anywhere else in the United States. Many of them are tropical species on the northern limit of their range in Florida, as are many of their native larval host plants and favorite nectar sources. Because the focus of this book is plants, I encourage readers to seek out books to help identify local butterflies. Check the Resources section in the back of this guide for recommendations.

A pair of white peacocks mating. The female will have only about ten days to find the proper larval host plant upon which to lay her eggs.

Larval Host Plants

"Well, I'll eat it," said Alice, "and if it makes me grow larger, I can reach the key; and if it makes me grow smaller, I can creep under the door: so either way I'll get into the garden, and I don't care which happens!"

Lewis Carroll, Alice's Adventures in Wonderland

In order to have a successful butterfly garden, you must understand the life cycle of butterflies. This knowledge is essential in your selection of plants and imperative when choosing larval host plants. There are two types of plants that butterflies visit, and these involve the two feeding stages of butterflies. One is a nectar source, which is a flowering plant that

Zebra longwing butterfly larvae on a passionflower plant.

adult butterflies visit to feed on flower nectar or, in some cases, pollen. The other is a source of food for the larval stage, which is a plant that the adult female butterfly lays eggs on and that the larvae eat until they are mature enough to pupate. Hungry butterfly larvae will eat leaves, flowers, stems, and sometimes even the fruits and seeds of their host plant, so expect this group of plants to be ravaged periodically by voracious feeding. Leaf damage from some species of butterfly larvae will hardly be noticeable, while others may strip the leaves and stems of the host plant down to the ground.

It is also useful to know that larval host plants typically are not unsightly weeds. Many larval host plants are as attractive as the flowering plants that adult butterflies visit for nectar. Some larval host plants that are utilized by common urban butterflies in tropical Florida include popular plants in cultivation, such as citrus (*Citrus* spp.), milkweeds (*Asclepias* spp.), and passionflowers (*Passiflora* spp.). Larval host plants can be kept in obscure areas of the garden if you prefer plants with chewed leaves to be less visible.

Mud in Your Garden

Many butterflies sip mineral-rich water from mud, an activity called *puddling,* and this is especially true of swallowtails, fritillaries, and large species of sulphurs. To supply them with puddling opportunities, simply take

A red admiral sips mineral-rich moisture from moist soil.

a birdbath, fill it with soil, and run your garden hose in it to create a mud hole for butterflies. It also makes for an interesting conversation piece to explain to your houseguests.

The Butterfly Life Cycle

The butterfly life cycle, comprising the development from egg to winged, adult butterfly, can be divided into four distinct stages:

1. Egg
2. Larva (or caterpillar)
3. Pupa (or chrysalis)
4. Adult butterfly

Eggs are generally laid on new leaves, petioles (leaf stems), or other parts of the larval host plant by the adult female butterfly. How many eggs are laid and how long it will take for the eggs to hatch depends on the species of

Eggs of the rare pink-spot sulphur cover the new growth on its larval host plant, horseflesh mahogany *(Lysiloma sabicu).*

butterfly. This is a very vulnerable stage in a butterfly's life because predators such as ants can attack the eggs.

Butterfly eggs typically hatch 3–6 days after they are laid. Once the eggs hatch, each larva undergoes growth stages called *instars,* and the average number of molts the larva undergoes before transforming into a pupa, or chrysalis, is five. Butterfly larvae have been described as being nothing more than a mouth attached to a long digestive tract. Butterfly larvae begin feeding at birth and are veritable eating machines, hardly pausing to rest as they feed throughout each day. The larvae of most species of butterflies feed as solitary individuals, but some species are gregarious and feed together in groups. Larvae of the Polydamas swallowtail (*Battus polydamas*) aggregate when young but become solitary with age. Feeding in groups could be a survival mechanism (safety in numbers), or it may simply be easier for young larvae to chew into tough leaves as a group rather than singly.

Some butterfly larvae feed on plants that contain toxic compounds. Some deal with this by detoxifying the compounds and passing them

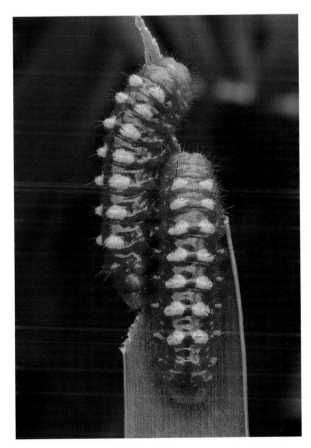

The garish colors of Atala butterfly larvae advertise their toxicity to would-be predators.

through their digestive tract, while others store the compounds in their own tissues for defense against predators. Black swallowtail larvae eat water hemlock with impunity, yet any small portion of the plant can kill an adult human. The larvae of monarch and queen butterflies typically feed on plants in the genus *Asclepias*, the milkweeds. Milkweeds contain heart toxins (cardiotoxic glycosides) that can be fatal to grazing livestock, yet the butterfly larvae eat them without ill effect. The distastefulness and toxicity of the plant is transferred to the butterfly larvae, making them distasteful and potentially toxic to birds and other predators. If a bird attacks a larva of a monarch butterfly, it will probably kill the larva. However, the resulting bad taste or digestive upset experienced by the bird teaches it to avoid similar-looking butterflies in the future. The milkweed toxins didn't help that particular larva from being killed by the bird, but they helped monarch butterflies as a whole. The next time that bird encounters a monarch larva it will likely leave it alone. This learning experience is often passed on to the bird's offspring as well.

Some butterflies have even evolved to resemble unpalatable species or objects in order to avoid being attacked by predators. This ploy is called *protective mimicry*. Giant swallowtail larvae mimic bird droppings to keep hungry birds from giving them a second thought. Most butterfly larvae, however, rely upon camouflage or simply good luck to keep from being eaten.

The time spent in the larval stage varies among species, but once a larva has molted into the fifth instar stage, it is ready to pupate. Some butterfly larvae will pupate directly on their host plant, like the Atala butterfly often does, but most species crawl away from their host plant and pupate elsewhere, which is typical of the monarch.

The final molt of a larva results in a resting stage called the pupa, or chrysalis (a special term used for butterfly pupae; the cases of moths are called *cocoons*). Again, the time spent as a pupa is dependent upon the butterfly species; this time span could involve a matter of days or many months. Some species overwinter as pupae, hatching out in spring, while others hatch during any month of the year.

Pupae are attached to natural or manmade objects by a series of hooks (called the *cremaster*) that are in the silk spun by the prepupal larva. This is often the only point of attachment, but some butterfly larvae suspend the pupa with additional strands of silk to help make it more secure.

Like eggs, pupae are vulnerable to predation, injury, and diseases. Migrating warblers, flycatchers, and vireos consume adult butterflies, larvae, and pupae, so the butterfly that manages to hatch and live a full natural

Pink-spot sulphur larvae rely on camouflage to avoid predation.

life is a lucky one. As metamorphosis advances, the pupa begins to change color, usually darkening, prior to the emergence of the adult butterfly. It is interesting to note that the pheromones produced by female butterflies are recognizable by males even in the pupal stage. It is not uncommon to find several adult male butterflies hovering around a pupa containing a female butterfly in anticipation of her emergence. Mating sometimes takes place before the female is fully emerged.

Monarch butterfly pupae resemble jeweled earrings.

Butterflies typically emerge from the pupal case in midmorning to take advantage of the warming and drying effects of the sun. Once fully emerged, the adult butterfly will poise quietly as its wings expand and dry, before taking flight. The average life span of an adult butterfly is less than two weeks (with some notable exceptions), so there is little time for a female butterfly to feed, mate, locate the appropriate larval host plant, and lay her own eggs to continue the cycle and perpetuate the species. This is why it is critically important for you, the gardener, to supply the correct nectar plants as well as larval host plants for species of butterflies that frequent your neighborhood. Without the correct larval host plants, you will not be able to maintain a resident breeding butterfly population. With the wholesale destruction of natural habitats in Florida, urban butterfly gardens become critically important to the survival of butterflies (plus many species of birds and even moths). Assess the various species of butterflies that occur in your neighborhood, and plant accordingly.

Monarchs Need Your Help

In 2013 an article in the *New York Times* reported that there has been a drastic decline in migrating monarch butterfly populations due to a number of causes, including the disappearance of native milkweeds in this country. This decline has been attributed to the use of genetically modified corn and other crops that are immune to herbicides, so now the entire crop can be sprayed with herbicide to eliminate competitive weeds without harming the crop. Milkweeds are collateral damage. The *New York Times* estimated that this practice has already eliminated about 120 million acres of monarch butterfly breeding habitat in the United States.

A pair of monarch butterfly larvae eat stems of a cultivated milkweed.

Gardeners can help reverse some of this loss by planting native and select exotic milkweeds. Gardeners in South Florida are lucky because we have a resident, nonmigratory population of monarchs to adorn our gardens all year long.

A Word about Butterfly Predators

While butterflies seek out nectar and pollen as a food source, there are other creatures that seek out butterflies for the same purpose. Warblers, vireos, gnatcatchers, flycatchers, mockingbirds, orioles, tanagers, and many other birds sometimes eat adult butterflies or their larvae and pupae. This is the natural order of things. It's simply a part of nature, so learn to enjoy the birds as well as the butterflies.

Lizards also commonly prey on butterflies and their larvae, as do tree frogs and even toads. Encouraging racers and cornsnakes will help keep

A black swallowtail pupa blends in with its surroundings to avoid detection by predators.

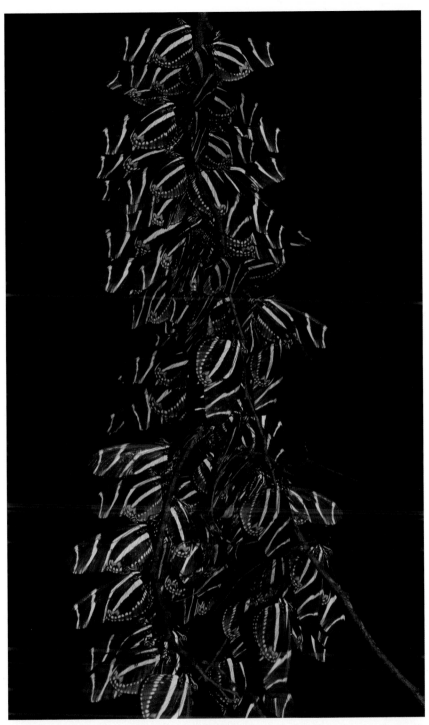

A roost of more than seventy zebra longwings in the author's garden. Communal roosting at night may be a tactic to avoid nocturnal predators.

lizards in check; cattle egrets, loggerhead shrikes, and kestrels also feast on a good share of lizards. The poisonous, non-native marine toad or giant toad should be humanely dispatched, especially if you own dogs.

Spiders claim a number of adult butterflies, both by capturing them in webs and through pollinator predation, which simply involves waiting patiently at flowers and nabbing whatever bug stops by for lunch. Ambush bugs, aptly named, are adept at pollinator predation and so are praying mantids. This is all a part of nature's grand theme of things, so no action is recommended on the part of the butterfly gardener.

The one creature you should control in and around butterfly gardens is the imported fire ant, because it attacks butterfly eggs and larvae (and

Giant swallowtail larvae avoid predation by mimicking bird droppings.

Fake eyespots on the hind wings of black swallowtails divert hungry birds away from the butterfly's body.

gardeners!). Fire ants make mounds of soil with interconnected tunnels from one mound to the next and can produce megacolonies that cover acres of land. A short-term method of dealing with fire ants is to pour a pot of boiling water on their mounds. The long-term control method is to use a biological control product sold under various trade names in most garden centers. Sprinkle the granules around the ant nests, and the workers will eventually carry them to the queen. Repeated treatments may be necessary to eradicate them. Try your absolute best to never use insecticides, especially on plants in your butterfly garden.

Butterflies

Many butterflies are easy to recognize, such as the zebra longwing, while others may look enough alike to confuse beginning and seasoned gardeners alike. Learning to distinguish between a large orange sulphur and an orange-barred sulphur may take some time, but don't get discouraged. You will soon gain confidence as you become accustomed to their field markings and see them over and over again. Other butterflies defy identification even by lepidopterists. The ones that cause the most mental anguish are the skippers. These are those little brown butterflies that fly fast—if one can categorize them that simply. The hairstreaks are another group that can be frustrating to identify. Although some groups are challenging, many people simply rejoice in butterflies being in and around their garden and do not really have a desire to name every species. If you do

The easily recognized zebra longwing is Florida's official state butterfly.

have a profound curiosity about the names of things, then there is help to be had in the plethora of books on Florida butterflies. There are even books that help identify butterfly larvae and pupae. Check the Suggested References section in the back of this guide for recommendations.

Being knowledgeable about which butterflies frequent your neighborhood will make you a better gardener, especially when it comes to adding larval host plants, because some butterflies are very host specific. Without proper host plants, they will not take up permanent residence in your garden.

One excellent way to learn about butterflies is to participate in guided butterfly walks. There are local chapters of the North American Butterfly Association (NABA) throughout Florida, and more are being formed as interest grows. There are also "butterfly counts" sponsored in many areas of Florida; these are wonderful activities to attend for both beginning and seasoned butterfly enthusiasts. You will not only meet the experts but also have the chance to develop new friends who have a shared interest. At NABA activities, experts are on hand to help you with your butterfly identifying skills; also, you will discover new places where you can watch butterflies on your own. Butterfly counts are fun, educational, and rewarding, and they also serve as a means to study butterfly population dynamics. You will find that we know surprisingly little about butterflies compared to birds. Butterfly counts are an essential way for scientists to gain valuable information about seasonal abundance, habitats, and behavior. Butterflies are also sensitive indicators of environmental health, so butterfly counts document population trends and help indicate whether or not a particular habitat is healthy or not. This vital information may lead to new resource-management policies, trigger more in-depth studies, or even initiate captive breeding programs for species on the brink of collapse.

To aid in observing butterflies, new binoculars have been designed specifically for butterfly watching. "Butterflying" is a growing pastime for many people and is beginning to rival the popularity of bird-watching in some circles. These binoculars focus at close range with a narrow field of vision. As you get more involved with butterflies and butterfly gardening, you will likely want to invest in a good pair of binoculars to use in your home garden and in the field. Close-focusing binoculars excel for bird-watching, too. Some of the best close-focusing binoculars are pricey, but their superior optics make them well worth the expense. Remember that investing in a good pair of binoculars will offer you a lifetime of joy

A hammock skipper has bold white spots on the forewings.

and pleasure, and they can be handed down to future butterfly- and bird-watchers.

For more information about the North American Butterfly Association, and for books to help with your identifying skills, see the Resources section in the back of this book.

Acknowledgments

First I must convey loving affection to my wife, playmate, and partner in pulling weeds, Michelle. She is a gardener at heart as she tends to her shade house brimming with exotic orchids and our acre of flowering plants, fruit trees, vegetables, and whatever else is out there. Seeing her eyes light up when hummingbirds and butterflies visit flowers in our garden is worth whatever effort it takes to dig planting holes in solid limestone.

I am forever grateful to my grandparents, Clarence and Olive Postle, and my mother, Martha, for teaching me about gardening and flowers at a very young age, which also rubbed off on my older brother, Russell. They influenced our lives more than they could possibly ever know.

The gardeners of Florida, myself included, owe a special debt to the many plant explorers—David Fairchild, Henry Nehrling, and Wilson Popenoe among them—who ventured throughout the world to discover new plants to cultivate. By foraging through remote, wonderful, and sometimes dangerous places, they have rewarded us all with their collections and helped define the charm and beauty of Florida's public and private gardens.

To the members of the Florida Native Plant Society, Tropical Audubon Society, Tropical Flowering Tree Society, North American Butterfly Association, and the many garden clubs of Florida, thank you for your genuine friendship and shared interests. Equal gratitude goes to the cheerful staff and volunteers at Fairchild Tropical Botanic Garden in Coral Gables for their friendly smiles as well as the countless plants and rich memories that I have acquired from plant sales. I cherish those memories even more than the plants. Gardens filled with flowering plants are also places of refuge for the human spirit, so I suppose I should even thank the plants in my own garden for giving me such pleasure and purpose in life.

Special gratitude goes to my lepidopterist friend Marc Minno and his charming wife, Maria, who reviewed and commented on the draft manuscript. Marc very capably made the butterfly list for tropical Florida more accurate than it otherwise would have been and he assisted in bringing the Latin names of butterflies up to date. Much appreciation also goes to my friend and colleague Craig Huegel for offering comments on the draft manuscript to help make this a better guide. I also thoroughly enjoyed the many e-mails sent back and forth to my dear friend Kirsten Llamas as we discussed some of the plants included in this guide. Her own book on tropical plants set a high standard for future writers on the subject. I am very much indebted to Doug Ingram, his sons William and Rodney, and his grandson Robbie, who operate Doug Ingram and Sons Nursery near Homestead, Florida. Having free access to roam their wholesale nursery in a golf cart with my camera made photographing some of the plants in this guide so easy it almost felt like cheating.

Foremost among the helpful staff and volunteers at Fairchild Tropical Botanic Garden in Coral Gables were Mary Collins, Marilyn Griffiths, David Hardy, Jason Lopez, Tiffany Lum, and Marlon Rumble. Other gardening and horticulturist friends who deserve an honorable mention include Maryanne Biggar, Ken Cook, Cindy David, Jesse Durko, Lisa Hammer, John Lawson, Richard Lyons, Stan Matthews, and Leslie Veber.

And heaping wheelbarrow loads of appreciation go to Meredith Babb and the rest of the staff at the University Press of Florida in putting this book together for the present and future gardeners in tropical Florida. It has been a humbling task.

Appendix

Checklist of Tropical Florida Butterflies

Swallowtails

☐	Pipevine swallowtail	*Battus philenor*
☐	Polydamas swallowtail	*Battus polydamas*
☐	Zebra swallowtail	*Eurytides marcellus*
☐	Bahamian swallowtail	*Heraclides andraemon*
☐	Schaus' swallowtail	*Heraclides aristodemus*
☐	Giant swallowtail	*Heraclides cresphontes*
☐	Black swallowtail	*Heraclides polyxenes*
☐	Eastern tiger swallowtail	*Pterourus glaucus*
☐	Palamedes swallowtail	*Pterourus palamedes*
☐	Spicebush swallowtail	*Pterourus troilus*

Whites and Sulphurs

☐	Pink-spot sulphur	*Aphrissa neleis*
☐	Statira sulphur	*Aphrissa statira*
☐	Florida white	*Appias drusilla*
☐	Great southern white	*Ascia monuste*
☐	Orange sulphur	*Colias eurytheme*
☐	Barred yellow	*Eurema daira*
☐	Dina yellow	*Eurema dina*
☐	Little yellow	*Eurema lisa*
☐	Sleepy orange	*Eurema nicippe*
☐	Mimosa yellow	*Eurema nise*
☐	Lyside sulphur	*Kricogonia lyside*
☐	Dainty sulphur	*Nathalis iole*
☐	Large orange sulphur	*Phoebis agarithe*
☐	Orange-barred sulphur	*Phoebis philea*

☐	Cloudless sulphur	*Phoebis sennae*
☐	Checkered white	*Pontia protodice*
☐	Southern dogface	*Zerene cesonia*

Hairstreaks

☐	Great purple hairstreak	*Atlides halesus*
☐	Red-banded hairstreak	*Calycopis cecrops*
☐	Amethyst hairstreak	*Chlorostrymon maesites*
☐	Silver-banded hairstreak	*Chlorostrymon simaethis*
☐	Fulvous hairstreak	*Electrostrymon angelia*
☐	Atala	*Eumaeus atala*
☐	Southern hairstreak	*Fixsenia favonius*
☐	Gray ministreak	*Ministrymon azia*
☐	White M hairstreak	*Parrhasius m-album*
☐	Bartram's scrub-hairstreak	*Strymon acis*
☐	Mallow scrub-hairstreak	*Strymon istapa*
☐	Disguised scrub-hairstreak	*Strymon limenia*
☐	Martial scrub-hairstreak	*Strymon martialis*
☐	Gray hairstreak	*Strymon melinus*

Blues

☐	Eastern pygmy-blue	*Brephidium isophthalma*
☐	Nickerbean blue	*Hemiargus ammon*
☐	Ceraunus blue	*Hemiargus ceraunus*
☐	Miami blue	*Hemiargus thomasi*
☐	Cassius blue	*Leptotes cassius*

Metalmarks

☐	Little metalmark	*Calephelis virginiensis*

Snout Butterflies

☐	American snout	*Libytheana carinenta*

Heliconians

☐	Gulf fritillary	*Agraulis vanillae*
☐	Julia	*Dryas iulia*
☐	Zebra longwing	*Heliconius charithonia*

Brushfoots

☐	White Peacock	*Anartia jatrophae*
☐	Cuban crescent	*Anthanassa frisia*
☐	Variegated fritillary	*Euptoieta claudia*
☐	Common buckeye	*Junonia coenia*
☐	Mangrove buckeye	*Junonia evarete*
☐	Tropical buckeye	*Junonia genoveva*
☐	Mourning cloak	*Nymphalis antiopa*
☐	Phaon crescent	*Phyciodes phaon*
☐	Pearl crescent	*Phyciodes tharos*
☐	Eastern comma	*Polygonia comma*
☐	Malachite	*Siproeta stelenes*
☐	Red admiral	*Vanessa atalanta*
☐	Painted lady	*Vanessa cardui*
☐	American lady	*Vanessa virginiensis*

Admirals

☐	Viceroy	*Basilarchis archippus*
☐	Dingy purplewing	*Eunica monima*
☐	Florida purplewing	*Eunica tatila*
☐	Ruddy daggerwing	*Marpesia petreus*

Leafwing Butterflies

☐	Florida leafwing	*Anaea floridalis*

Emperors

☐	Hackberry emperor	*Asterocampa celtis*
☐	Tawny emperor	*Asterocampa clyton*

Milkweed Butterflies

☐	Soldier	*Danaus eresimus*
☐	Queen	*Danaus gilippus*
☐	Monarch	*Danaus plexippus*

Satyrs

☐	Carolina satyr	*Hermeuptychia sosybius*
☐	Georgia satyr	*Neonympha areolata*

Spread-Winged Skippers

☐	Silver-spotted skipper	*Epargyreus clarus*
☐	Zestos skipper	*Epargyreus zestos*
☐	Florida duskywing	*Ephyriades brunneus*
☐	Sleepy duskywing	*Erynnis brizo*
☐	Horace's duskywing	*Erynnis horatius*
☐	Juvenal's duskywing	*Erynnis juvenalis*
☐	Zarucco duskywing	*Erynnis zarucco*
☐	Mangrove skipper	*Phocides pigmalion*
☐	Hammock skipper	*Polygonus leo*
☐	Common checkered-skipper	*Pyrgus communis*
☐	Tropical checkered-skipper	*Pyrgus oileus*
☐	Hayhurst's scallopwing	*Staphylus hayhurstii*
☐	Southern cloudywing	*Thorybes bathyllus*
☐	Confused cloudywing	*Thorybes confusis*
☐	Northern cloudywing	*Thorybes pylades*
☐	Dorantes longtail	*Urbanus dorantes*
☐	Long-tailed skipper	*Urbanus proteus*

Grass Skippers

☐	Dusky roadside-skipper	*Amblyscirtes alternata*
☐	Delaware skipper	*Anatrytone logan*
☐	Least skipper	*Ancyloxypha numitor*
☐	Monk skipper	*Asbolis capucinus*
☐	Sachem	*Atalopedes campestris*
☐	Dusted skipper	*Atrytonopsis hianna*
☐	Brazilian skipper	*Calpodes ethlius*
☐	Southern skipperling	*Copaeodes minimus*
☐	Three-spotted skipper	*Cymaenes tripunctus*
☐	Palmetto skipper	*Euphyes arpa*
☐	Berry's skipper	*Euphyes berryi*
☐	Palatka skipper	*Euphyes pilatka*
☐	Dun skipper	*Euphyes vestris*
☐	Dotted skipper	*Hesperia attalus*
☐	Meske's skipper	*Hesperia meskei*
☐	Fiery skipper	*Hylephila phyleus*
☐	Clouded skipper	*Lerema accius*
☐	Eufala skipper	*Lerodea eufala*
☐	Swarthy skipper	*Nastra lherminier*

☐	Neamathla skipper	*Nastra neamathla*
☐	Twin-spot skipper	*Oligoria maculate*
☐	Ocola skipper	*Panoquina ocola*
☐	Salt marsh skipper	*Panoquina panoquin*
☐	Obscure skipper	*Panoquina panoquinoides*
☐	Baracoa skipper	*Polites baracoa*
☐	Tawny-edged skipper	*Polites themistocles*
☐	Whirlabout	*Polites vibex*
☐	Byssus skipper	*Problema byssus*
☐	Southern broken dash	*Wallengrenia otho*

Giant-Skippers

☐	Yucca giant-skipper	*Megathymus yuccae*

Resources for Gardeners

Places of Inspiration

Bok Tower Gardens
    1151 Tower Road
    Lake Wales, FL 33853

Fairchild Tropical Botanic Garden
    11935 Old Cutler Road
    Coral Gables, FL 33156

Flamingo Gardens
    3750 South Flamingo Road
    Davie, FL 33330

Florida Botanical Gardens
    12520 Ulmerton Road
    Largo, FL 33774

Heathcote Botanical Gardens
    210 Savannah Road
    Fort Pierce, FL 34982

The Kampong
    4013 South Douglas Road
    Miami, FL 33133

Key West Botanical Garden
    5210 College Road
    Key West, FL 33040

Marie Selby Botanical Gardens
    811 South Palm Avenue
    Sarasota, FL 34236

McKee Botanical Garden
    350 U.S. Highway 1
    Vero Beach, FL 32962

Mounts Botanical Garden
    531 North Military Trail
    West Palm Beach, FL 33415

Naples Botanical Garden
    4820 Bayshore Drive
    Naples, FL 34112

Redland Fruit and Spice Park
    24801 Southwest 187 Avenue
    Homestead, FL 33031

Societies

Florida Native Plant Society

Everyone reading this book should join the Florida Native Plant Society. There are local chapters throughout Florida that hold monthly meetings, and these are a good source of information as well as plants grown by members for the raffle table. Local chapters offer educational programs and guided field trips, and they often sponsor visits to home gardens featuring Florida native plants in the landscape. The annual state conference is also a wealth of information, and vendors there sell plants that are difficult to find elsewhere. Visit their Web site for further information on local chapters and how to join the organization.

National Audubon Society

When you join the National Audubon Society, you automatically become a member of the Florida Audubon Society as well as your local Audubon Society chapter. Audubon Society membership allows free entry into Corkscrew Swamp Sanctuary, and local chapters offer guided field trips into various parks and preserves throughout the year. They sometimes even sponsor out-of-country trips. Visit the Florida Audubon Society Web site to find the chapter nearest you.

North American Butterfly Association (NABA)

This is another organization you should definitely join. They offer educational programs, guided field trips, and regional butterfly counts, plus some chapters sponsor visits to butterfly gardens established in members' own yards. NABA field trips are wonderful opportunities to meet people with shared interests, discover new places to visit on your own, and learn how to identify butterflies from the experts. Visit the North American Butterfly Association Web site to find the chapter nearest you.

Tropical Flowering Tree Society

This group sponsors plant sales in South Florida that are a valuable source of very difficult-to-find plants. If you are a collector of unusual plants, then this is a good society to join even though nonmembers can still attend the plant sales.

Parks and Preserves

Many outstanding public parks and preserves occur within the range of this book and offer viewing of native plants, butterflies, hummingbirds, and other wildlife in their natural habitats. Visit their Web sites for detailed information on individual parks and preserves located in the area where you live or where you plan to visit.

National Parks and Preserves

Arthur R. Marshall Loxahatchee National Wildlife Refuge
Big Cypress National Preserve
Biscayne National Park
Everglades National Park
Merritt Island National Wildlife Refuge
National Key Deer Refuge

State Parks and Preserves

For a comprehensive listing of Florida's award-winning state parks and preserves, visit the Florida State Parks Web site.

County Parks and Preserves

For a comprehensive listing of county parks and preserves, visit the Web sites of individual counties.

Other Parks, Preserves, and Attractions

Butterfly World
Located in Tradewinds Park/Broward County
3600 Sample Road
Coconut Creek, FL 33073

Butterfly World is a wonderful source of unusual plants to attract butterflies and is a wonderful and fun educational experience. Kids love it.

Corkscrew Regional Ecosystem Watershed (CREW Marsh)
23998 Corkscrew Road
Estero, FL 33928

Guided wildflower walks and butterfly walks are offered during the year, and the 60,000-acre preserve is an outstanding place to bird-watch, hike, bike, or even camp.

Corkscrew Swamp Sanctuary
375 Sanctuary Road West
Naples, FL 34120

This outstanding National Audubon Society sanctuary offers a 2½-mile elevated boardwalk that allows you to visit a swamp without getting your feet wet. It is a premier birding and butterflying destination. An Audubon Society membership gets you in free.

Organic Pest Controls

If you need to control pests in and around your hummingbird and butterfly garden (or vegetable garden), a wide array of organic products and natural controls are available to the environmentally conscious gardener. Here is one of the best sources:

Gardens Alive!
5100 Schenley Place
Lawrenceburg, IN 47025

This mail-order company offers award-winning products that will help keep your garden (and you!) healthy without harmful chemicals. They can also provide live lady beetles and other beneficial insects. Their catalog also offers expert advice on how to prevent and combat garden problems and is a must-have for all gardeners.

Florida Native Wildflower Seeds
Florida Wildflower Growers Cooperative
   (provider of Florida wildflower seed packets)

Learn More About Florida Native Wildflowers
Florida Wildflower Foundation
225 S. Swoope Avenue, Suite 110, Maitland, FL 32751

University of Florida Master Gardener Program

If you wish to learn more about gardening, sharpen your horticulture skills, and become active in your community as a volunteer, the Florida Master Gardener Program is for you. This program benefits the University of Florida's Institute of Food and Agricultural Sciences Extension as well as the citizens of Florida. For further information, contact your local Agricultural Extension Service office or visit the program's Web site.

Certify Your Yard

You can have your property designated a Certified Wildlife Habitat by several organizations, but the National Wildlife Federation is one of the best. You will need to fill out a form and supply photographs that show you have met the criteria for certification. These include providing food, water, cover, and places to raise young through the use of native plants. It can also include providing nest boxes, water features, and butterfly gardens. Visit the National Wildlife Federation's Backyard Wildlife Certification Web site for details.

Selected Bibliography

Cech, Rick, and Guy Tudor. *Butterflies of the East Coast.* Princeton: Princeton University Press, 2007.

Daniels, Jaret C. *Your Florida Guide to Butterfly Gardening.* Gainesville: University Press of Florida, 2000.

Glassberg, Jefrey, Marc C. Minno, and John V. Calhoun. *Butterflies Through Binoculars: A Field, Finding, and Gardening Guide to Butterflies in Florida.* New York: Oxford University Press, 2000.

Hammer, Roger L. *Everglades Wildflowers.* Guilford, Conn.: Globe Pequot Press, 2002.

———. *Florida Keys Wildflowers.* Guilford, Conn.: Globe Pequot Press, 2004.

Howell, Steve N. G. *Hummingbirds of North America: The Photographic Guide.* Princeton: Princeton University Press, 2003.

Huegel, Craig N. *Native Plant Landscaping for Florida Wildlife.* Gainesville: University Press of Florida, 2010.

———. *Native Wildflowers and Other Ground Covers for Florida Landscapes.* Gainesville: University Press of Florida, 2012.

Johnsgard, Paul A. *The Hummingbirds of North America.* 2nd ed. Washington, D.C.: Smithsonian Institution Press, 1997.

Llamas, Kirsten Albrecht. *Tropical Flowering Plants.* Portland, Ore., 2003.

Minno, Marc C., Jerry F. Butler, and Donald W. Hall. *Florida Butterfly Caterpillars and Their Host Plants.* Gainesville: University Press of Florida, 2005.

Minno, Marc C., and Thomas Emmel. *Butterflies of the Florida Keys.* Gainesville: University Press of Florida, 1993.

Minno, Marc C., and Maria Minno. *Florida Butterfly Gardening.* Gainesville: University Press of Florida, 1999.

Nelson, Gil. *Florida's Best Native Landscape Plants.* Gainesville: University Press of Florida, 2003.

Osorio, Rufino. *A Gardener's Guide to Florida's Native Plants*. Gainesville: University Press of Florida, 2001.

Peterson, Roger Tory. *Birds of Eastern and Central North America*. 6th ed. Boston: Houghton Mifflin Harcourt, 2010.

Read, Robert W. *Nehrling's Early Florida Gardens*. Gainesville: University Press of Florida, 2001.

Taylor, Walter Kingsley. *Florida Wildflowers: A Comprehensive Guide*. Gainesville: University Press of Florida, 2013.

Vanderplank, John. *Passion Flowers*. 3rd ed. Cambridge, Mass.: MIT Press, 2000.

Index of Plant Names

Page numbers in italics indicate illustrations.

Holmskioldia tettensis, 86
Honeysuckle, Cape, 140, *140*
Honeysuckle, Coral, 157, *157*
Honeysuckle, Mexican, 85, *85*

Indiangrass, Lopsided, 212, *212*
Indigoberry, White, 115, *115*
Indigobush, Southern, 45, *45*
Iochroma cyaneum, 79, *79*
Ipomoea batatas, 156
Ipomoea cordatotriloba, 156
Ipomoea horsfalliae, 155, *155*
Ipomoea microdactyla, 156, *156*

Jack-in-the-Bush, 63, *63*
Jatropha integerrima, *23, 35, 35*
Jealousy, Maiden's, 167, *167*
Juanulloa aurantiaca, 80, *80*
Justicia, Summer Sun, 84, *84*
Justicia brandegeana, 81, *81*
Justicia brasiliana, 82, *82*
Justicia carnea, 83, *83*
Justicia corumbensis, 84, *84*
Justicia leonardii, 84
Justicia spicigera, 85, *85*

Karomia tettensis, 86, *86*
Kleinia fulgens, 196, *196*
Koanophyllon villosum, 87, *87*
Kosteletzkya pentacarpos, 88, *88*
Kosteletzkya virginica, 88

Lantana, Rockland, 197, *197*
Lantana camara, 11, *11*, 197, 198
Lantana depressa var. *depressa*, 12, 197, *197*
Lantana involucrata, 12, 89, *89*
Lantana 'Lola', 198, *198*
Lantana × *strigocamara*, 198
Lavender, Sea, 53, *53*
Leadplant, Crenulate, 45
Leaf, Vanilla, 181
Leonotis leonurus, 42, *42*
Leonotis leonurus, 90, *90*
Leopard Plant, Green, 189, *189*
Liatris elegans, 199
Liatris spicata, 199

Liatris spp., 199, *199*
Lignumvitae, 32, *32*
Lilac, Summer, 56, *56*
Lime, Wild, 41, *41*
Lippia nodiflora, 203
Lips, Blue, 129, *129*
Live Oak, 21, *21*
Lobelia cardinalis, 200, *200*
Lola, 198, *198*
Lonicera sempervirens, 157, *157*
Loosestrife, Winged, 92, *92*
Lycium carolinianum, 91, *91*
Lysiloma latisiliquum, 20, *20*
Lysiloma sabicu, 21, *21*, 230
Lythrum alatum var. *lanceolatum*, 92, *92*

Mahogany, Horseflesh, 21, *21, 230*
Mallow, Mangrove, 36
Malvaviscus arboreus var. *drummondii*, 93, *93*
Malvaviscus penduliflorus, 94, *94*
Man-in-the-Ground, 156, *156*
Manjack, Bahama, 67, *67*
Maple, Flowering, 43, *43*
Maypop, 160, *160*
Melanthera nivea, 95, *95*
Melanthera parvifolia, 95
Melochia spicata, 96
Melochia tomentosa, 96, *96*
Mikania scandens, 158, *158*
Milkpea, Eastern, 154, *154*
Milkpea, Elliott's, 154
Milkweed, Butterfly, 176, *176*
Milkweed, Giant, 61, *61*
Milkweed, Prairie, 175
Milkweed, Rose, 175, *175*
Milkweed, Scarlet, 174, *174*
Milkweed, Swamp, 175
Mimosa, Sunshine, 201, *201*
Mimosa quadrivalvis, 201
Mimosa strigillosa, 201, *201*
Mistflower, Blue, 184, *184*
Monarda didyma, 97, *97*
Monarda punctata, 98, *98*
Morning-Glory, Lady Doorly's, 155, *155*
Mucuna bennettii, 148
Musa ornata, 99, *99*

Roger L. Hammer is a professional naturalist, now retired. He received the first Marjory Stoneman Douglas Award presented by the Dade Chapter of the Florida Native Plant Society in 1982 for "outstanding, consistent, and constant service in the areas of education, research, promotion, and preservation of native plants." Tropical Audubon Society awarded him the prestigious Charles Brookfield Medal in 1996 for "outstanding service in the protection of our natural resources," and in 2003 he received the Green Palmetto Award in Education from the Florida Native Plant Society. In 2008 he gave the keynote address at the 19th World Orchid Conference in Miami and was a keynote speaker at the 2013 Florida Native Plant Society's state conference in Jacksonville, Florida. In 2012 he received an honorary Doctor of Science degree from Florida International University. Roger's hobbies include gardening, long-distance canoeing, camping, kayak fishing, birding, and wildflower photography. Roger is the author of *Everglades Wildflowers*, *Florida Keys Wildflowers*, the *Falcon Guide to Everglades National Park and the Surrounding Area*, and *Florida Icons: 50 Classic Views of the Sunshine State*. He lives in Homestead, Florida, with his wife, Michelle.